Howard W. Sams & Company

Complete Guide to Audio

Howard W. Sams & Company

Complete Guide to Audio

By John J. Adams

PROMPT®
PUBLICATIONS

©1998 by Howard W. Sams & Company

PROMPT© Publications is an imprint of Howard W. Sams & Company, A Bell Atlantic Company, 2647 Waterfront Parkway, E. Dr., Indianapolis, IN 46214-2041.

All rights reserved. No part of this book shall be reproduced, stored in a retrieval system, or transmitted by any means, electronic, mechanical, photocopying, recording, or otherwise, without written permission from the publisher. No patent liability is assumed with respect to the use of the information contained herein. While every precaution has been taken in the preparation of this book, the author, the publisher or seller assumes no responsibility for errors or omissions. Neither is any liability assumed for damages resulting from the use of information contained herein.

International Standard Book Number: 0-7906-1128-7
Library of Congress Catalog Card Number: 97-68183

Acquisitions Editor: Candace Drake Hall
Editor: Natalie F. Harris
Assistant Editors: Pat Brady, Loretta Leisure
Typesetting: Natalie Harris
Indexing: Natalie Harris
Cover Design: Debra Delk, Phil Velikan
Graphics: John Adams
Graphics Conversion: Christy Pierce, Dave Pruett, Kelli Ternet, Terry Varvel
Additional Text: Jim Adams
Additional Illustrations and Other Materials: Courtesy of Denon Electronics, Lenbrook Industries, Marantz Canada, Gabe Martin, Sony of Canada Ltd.

Trademark Acknowledgments:
All product illustrations, product names and logos are trademarks of their respective manufacturers. All terms in this book that are known or suspected to be trademarks or services have been appropriately capitalized. PROMPT© Publications, Howard W. Sams & Company, and Bell Atlantic cannot attest to the accuracy of this information. Use of an illustration, term or logo in this book should not be regarded as affecting the validity of any trademark or service mark.

PRINTED IN THE UNITED STATES OF AMERICA

9 8 7 6 5 4 3 2 1

CONTENTS

Preface   1

Chapter 1
Introduction to Audio Equipment   5
    Three Elements of an Audio System   5
        Source   *6*
        Amplification   *7*
        Output (Speakers)   *7*
    High Fidelity (Hi-Fi)   7
    Stereo: A Brief Explanation   8
    Audio Equipment Ultra Basics   8
        Composition of Typical Stereo System   *9*
        Real-World Stereo and
            Surround Sound Systems   *10*
    Which System Should You Purchase?   12
    What Constitutes an Average Audio System?   13
    What Constitutes an Expanded Audio System?   14
    What Constitutes a Light Audio System?   14
    How You Can Improve Your Audio System   15
    25 Audio Terms You Should Know   16
    Wrap Up   20

Chapter 2
Sound Revealed   21
    Mediums, Source Points & Receipt Points   22
    Waves   22
        Sound Waves   *22*
    Frequency & Pitch   23

Amplitude	24
Phase	24
Speed of Sound	24
Decibels	25
What a Human Ear Can Hear	26
Acoustics	27
How Sound is Created in Audio Equipment	*27*
Acoustic Properties in Audio Systems	*28*
How Sound is Recorded	29
How Sound is Stored	30
How Sound is Played	30
Signal-to-Noise Ratio	30
Wrap Up	30

Chapter 3
Stereophonics Revealed   **33**

Channels	34
Life Before Stereo: Monophonic	34
The Basics of Stereophonics	34
How the Magic is Performed	*36*
Another Way Stereo is Achieved	*36*
Stereo Imaging	*37*
Reverberation Makes the Sound Real	*38*
Stereo Gets Complex	38
Surround Sound	*39*
Recreating Events	39
Wrap Up	40

Chapter 4
Home Theater Sound   **41**

The Sound Experience	41
Surround Sound Takes Over	42
How Surround Sound Differs From Normal Stereo	*42*
Surround Sound Composition	*43*
Dolby Surround Sound	*43*

Dolby Pro Logic Surround Sound	43
Dolby Digital: AC-3	44
How Surround Sound is Encoded	45
Equipment	45
Surround Sound Signal Source	45
Dolby Pro Logic Sources	46
Dolby Digital (AC-3) Sources	47
Receivers and Decoders	47
Speakers	48
Amplifiers	49
DTS and Other Non-Dolby Surround Sound Systems	49
Wrap Up	50

Chapter 5
Amplifiers & Preamplifiers — 51

How an Amplifier Works	52
Audio Amplifiers	52
Preamplifiers	*53*
Workings	53
How Preamps are Used in a Modern Stereo	54
Power (Output) Amplifiers	*54*
Workings	55
The Misconception of Watts	55
Various Types of Amplifiers	55
Vacuum Tube Amp	*56*
Solid-State Transistor Amp	*56*
What Kind of Amp do You Need?	56
Noise & Distortion	57
Channel Separation	57
Impedance Matching	57
How Manufacturers Measure the Wattage of a Power Amp	58
Examples	*58*
Wrap Up	59

Chapter 6
Receivers & Surround Sound Decoders   61
- The Tuner   61
  - *Radio Waves*   62
  - *How a Tuner Works*   63
- Receiving the Strongest, Least
  - Noisy FM Signal   64
  - *Antennas*   64
  - *Digital Audio Broadcast (DAB)*   65
- The Surround Sound Decoder   65
- DPSs   66
- Remote Controls   66
- Real-World Receivers   67
  - *What to Look For in a Receiver*   67
  - *What to Avoid*   67
- Future Receivers   69

Chapter 7
Cassette Decks, CD Players, DVD Players, MiniDisc & Phonographs   71
- Cassette Decks   72
  - *The Innards of a Tape Deck*   73
  - *How the Parts Interact*   73
  - *Types of Tape Decks*   74
- Digital Audio Tape: DAT   75
  - *What to Look for in a Tape Deck*   76
- Compact Disc Players   76
  - *Let's Get Digital*   76
  - *What are the Ones and Zeros?*   77
  - *Analog-to-Digital & Digital-to-Analog Converters*   78
  - *How a CD Changes Digital Information into Audio Waves*   79
  - *Sampling*   80
  - *Oversampling*   80
  - *CD Players*   80
  - *Digital Video (Versatile) Disc Player*   82

MiniDisc: MD	82
The Discs	*83*
ATRAC	*83*
MiniDisc Recorders	*84*
The Phonograph	84
Parts of a Record Player	*84*
How an LP Works	*84*
Phonograph Recommendations	*85*
Wrap Up	85

Chapter 8
Loudspeakers, Headphones & Microphones — 87

Loudspeakers	88
Getting it Right	*89*
Drivers	*89*
Composition of a Loudspeaker	90
Cabinet or Enclosures	*90*
Driver Types	*91*
Crossover Network	*92*
How a Loudspeaker Works	*93*
Common Loudspeakers	93
Conventional Front Set	*94*
Sub/Sat	*94*
Satellites	*95*
Subwoofers	*95*
Surround Sound Speakers	*96*
How to Set Up Your Speakers	96
Setting Up Surround Sound Speakers	*99*
Power Ratings	99
Digital Speaker Systems	99
Headphones	100
Why Headphones Sound Different From Speakers	*100*
Seals	*100*
What to Look For	*101*
Microphones	101
Microphone Mechanics	*102*
Two Types of Mics	*102*

    Directionality of a Microphone   *102*
    Choosing a Microphone   *102*
  Wrap Up   103

Chapter 9
Computer Sound   105
  Sound Cards   105
  16-Bit Sound   106
  MIDI   106
  3D Sound   107
  What Card to Buy   107
  Speakers   108
  Wrap Up   109

Chapter 10
Brands & Choices   111
  Questions to Ask Yourself Before Looking   111
  Steps to Deciding Brands and Models   112
    Research   *112*
  Brands   113
    How Does it Sound?   *113*
    Other Questions   *113*
    Brands to Look For   *114*
    Rating a Brand   *114*
  Rating How a System Sounds   114
  Where to Purchase   116
  Purchasing Strategies   117
    Other Purchasing Tips   *118*
  Recommendations   118
    Combination, Separate or Rack?   *118*
    System Strategies   *119*
    Receivers   *120*
    Loudspeakers   *120*
    CD Players   *121*
    Tape Decks   *121*
  Wrap Up   121

Chapter 11
Hookups & Accessories — **123**

- What You Need — 123
  - *Cables* — *123*
- Connectors — 125
- What Each Connection on Your Receiver is For — 126
  - *Typical Connections on an A/V Receiver* — *127*
- Hooking Up Components — 128
  - *Hooking Up Your Speakers* — *130*
  - *Hooking Up a Subwoofer* — *131*
  - *Hooking Up a CD Player to Your Receiver* — *131*
  - *Hooking Up a CD Player to Your Receiver* — *131*
  - *Hooking Up a Tape Deck to Your Receiver* — *131*
  - *Hooking Up a Turntable to Your Receiver* — *133*
  - *Hooking Up a VCR to Your Receiver* — *133*
  - *Hooking Up a Laserdisc Player to Your Receiver* — *135*
  - *Hooking Up a DVD Player to Your Receiver* — *135*
  - *Hooking Up External Amplifiers* — *136*
- Wrap Up — 136

Chapter 12
Features — **139**

- Receivers — 139
- Loudspeakers — 140
- CD Players/Changers/Jukeboxes — 141
- Tape Decks — 141
- DVD Players — 142

Appendix
Web Addresses **145**
    FAQs and Helpful Internet Links   147

Index **149**

*Dedicated to EVERYONE who brought out the writer in me.
It is appreciated. Thanks.*

PREFACE

Stereo equipment is needed to bring you the symphony of the ages on CD, or a bang-your-head rock tape. It may be used to set the evening scene with mood music or to record your child's first words. It brings vocal and instrumental masterpieces (or cacophonies, depending on your interpretation) into our homes, and permanently etches life memories into electronics.

The audio industry has matured along with the technology. Audio/visual equipment has all but replaced ancient two-speaker stereo systems with new multichannel surround sound. This creates a whole movie experience while still providing two-channel stereo for musical enjoyment. But why do we need a surround system that encircles our bodies with sound waves?

Imagine watching your favorite movie. The main character makes their way down a corridor towards a closed door. Eerie music fills the air. The character inches closer to the door and reaches out a hand… BLAM! The door is blown off its hinges with an ear-shattering explosion. All this action gets the heart pumping, but let's rewind and try an experiment. Turn the volume down all the way. Now watch the same scene. Not the same, is it? The sound is half the movie experience and needs to be treated that way, with quality sound equipment.

Complete Guide to Audio will answer questions you may have asked audio salespeople in the past but were unable to get an answer for. What equipment do you need? How is sound created and recreated? What makes up a basic/midrange/advanced sound system? Why are some brands cheaper than others? Will a CD player sound better than a MiniDisc? How does it all work?

With comprehensive, simple explanations, we will shed some light on sound. This book is written for you, as a consumer who is curious about the mystery behind those black boxes that pump tunes into our homes.

Ask yourself, are you an audiophile interested in the technical aspects of sound (equipment, frequencies, instruments, etc.)? Or do you simply want a good stereo to enjoy the feel and emotion of the music? In this book, we will balance these technical and subjective issues.

In addition, *Complete Guide to Audio* will explain some common problems that you may run into while setting up your stereo system and home entertainment center. Included are notes to answer those simple questions you have always had but never bothered to ask for fear of appearing uneducated (a consumer's worst mistake). We will strive to answer such questions as, "What exactly is Dolby?" or "How do I place my speakers?"

Chapter 1 will lay out a blueprint of the average sound system and the components you can chose from, as well as introduce you to stereo system basics and audio terms.

Chapter 2 gives you a firm background in the basics of sound — how it travels, how tones and loudness are determined, etc.

Chapter 3 delves into how stereo works, while home theater systems are explained in detail in chapter 4.

Chapters 5 through 9 will explain the separate components (amplifiers, receivers, CD players, tape decks, speakers, etc.) and how to make the best choices as well as define the best brands.

Computer sound is becoming a popular audio market, so chapter 9 will give it a rundown.

Move on to Chapter 10 for purchasing advice. Chapter 11 shows how to hook up components and what accessories are needed. Chapter 12 lists some common features found on audio equipment.

In acknowledgment, special thanks to Kristy Klein for her love and support. To Jim Adams for hours of help. To Pam Cassa, Dorothy Christiansen, Glen Klein, Bill Hatch, Missy, Amelia and Richard Garant. To Dashwood's and staff. Thanks to Katherine Clarke and Lenbrook Industries Limited, AKA Marantz Canada, for expediting photos. And to Natalie Harris, Candace Drake Hall and all the Prompt Publications staff (great people to work with). I also want to thank MPT Computers for saving my poor ailing computer at a key moment.

This book was written for the consumer and for people who want to know more about that mystical energy called sound. Hopefully this book will give you the necessary information to make successful purchases and demystify the wire spaghetti behind your system.

Enjoy!

John Adams

P.S. I am always happy to receive E-mail from readers:
Electronics@pobox.com
My personal Website is http://www.basicelectronics.com.

CHAPTER 1
INTRODUCTION TO AUDIO EQUIPMENT

"The wireless music box has no imaginable commercial value. Who would pay for a message sent to nobody in particular?"
David Sarnoff's associates in response to his urgings
for investment in the radio in the 1920s

In this chapter, we will take a peek at how audio equipment is the modern-day application of sound theory. In the next chapter, we will delve deeper into the mystical vibrations of sound and its idiosyncrasies. This may be putting the dessert before the main course, but learning some of the audio equipment basics at the outset is worth spoiling the sound dinner.

A modern hi-fi audio system creates sound by utilizing modern electronic ingredients. This concoction of transistors, resistors, capacitors and microprocessors fires electrons up and down paths of wires to bring us entertainment. What is the menu of sound equipment that will fill an acoustic appetite? How do the various ingredients of electronic components meld together to form sound? On with this hi-tech feast. Bon appetit!

THREE ELEMENTS OF AN AUDIO SYSTEM

An audio system is composed of equipment that provides an audio signal's source, amplification and output. See Figure 1-1.

Sound consists of air vibrations. It increases and decreases the air pressure sensed by our ears, face and chest. If you strike a snare drum, it pushes and pulls the surrounding air and transfers the vibes over a distance to our ears. Our brain picks up the slight motions on the eardrums, stimulating the nerves inside the ear. This we interpret as audio, or sounds. A modern audio system uses these same sound principles.

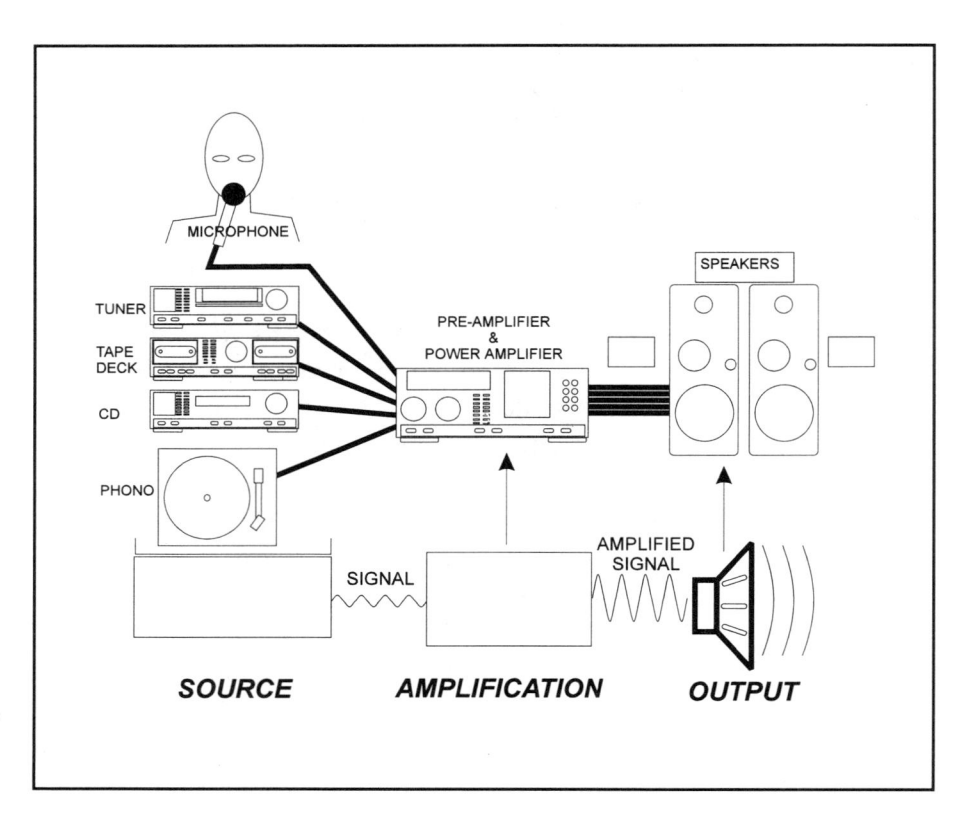

Figure 1-1.
Three elements
of an audio
system.

However, we are using equipment vastly more high-tech than a snare drum to reproduce the same vibrations. Here are the basic sections of audio equipment:

SOURCE

The point of origin for sound is known as the *source*. This is the spot in which the signal (energy) is first created. In our example of the snare drum, the source was the stick striking the surface. In audio equipment, this can be a microphone, tape deck, compact disc (CD) player, turntable or tuner (AM/FM). Each of these devices, with the exception of the microphone, is feeding the rest of the audio equipment prerecorded sounds that had their beginnings in a microphone setup. So the source can be a live signal (in the case of the microphone) or a facsimile (as with the CD, tape deck, etc.).

AMPLIFICATION

The volume coming from a microphone, CD, turntable, etc., is actually quite low. In fact, the level is so small that by itself, the energy would barely be enough to drive an ear piece. Therefore, in order to hear the sound, it must be *amplified* (increased) to audible levels. This is accomplished electronically with *audio amplifiers*.

Modern audio equipment makes use of two types of electronic amplifiers:

Preamplifier: Used to boost weak signals from turntable needles, microphones and other pieces of equipment for later use by the power amplifier. It also controls such things as volume, bass, treble, balance and equalization.

Power Amplifier or Output Amplifier: This amplifier gives sound the "punch" it needs to reach your ears at audible levels.

OUTPUT (SPEAKERS)

Once the sound signal is increased by the various amplification components of your stereo, it must be output. How else are we going to hear it? *Loudspeakers*, or more accurately, *drivers*, are electromechanical devices that *change* the electrical energy coming from the amplifier into a physical to-and-fro energy movement called *acoustic energy*.

Electrodynamic speakers are the most common type. They use a permanent magnet and electromagnet to drive a cone back and forth, thus displacing the surrounding air and creating sound. The harder and farther it pushes the cone, the louder the sound coming from the speaker.

HIGH FIDELITY (HI-FI)

What exactly is hi-fi? You will often hear the term with respect to audio equipment. But have you really defined it for yourself?

Fidelity is the accuracy with which a system, such as a stereo set, reproduces the original signal fed to it. *High fidelity* is the technique of recording, broadcasting and reproducing sound to match (as closely as possible) the characteristics of the original sound. The goal of hi-fi is to reproduce an identical set of sound vibrations. Have you ever listened to a pre-1940 radio recording? Most of the original sound vibrations are lost in the recording or playback on these "*lo-fi*" systems. Hi-fi, on the other hand, sounds *realistic*!

In order to achieve high fidelity, the recording and playback equipment must reproduce the sound virtually without distortion. It also must be able to span the entire human hearing frequency range (between 10 and 20,000 cycles per second, explained later in Chapter 2). This is possible with modern audio equipment.

STEREO: A BRIEF EXPLANATION

Stereophonic sound, or *stereo* for short, is a multichannel audio system. It means, literally, three-dimensional (3D) sound. Old *monophonic* systems had one-dimensional sound.

When we are dealing with a typical hi-fi system, we have two channels. This means the recording was coded with two separate tracks from two separate recording sources (microphones). These tracks are then played back through two speakers to recreate the stereo image that was created with the original recording.

Surround sound, on the other hand, uses three or more channels. In fact, the newest Dolby Digital surround sound systems use 6 channels. See Chapter 3 for more information on the magic of stereo, and Chapter 4 for surround sound systems and explanations.

AUDIO EQUIPMENT ULTRA BASICS

Now that you know the three elements of audio equipment, let's see what is really available in stores. What are the subtle and obvious differences? Let's get a feel for the audio market with these next sections.

We will discuss purchasing strategies and how to select a model in Chapter 10.

COMPOSITION OF TYPICAL STEREO SYSTEMS

Home stereo systems can be broken into two basic categories: stereo or surround sound. Each of these can be sheared into subcategories. Home stereo and surround components are virtually the same with the exception of the special receiver/decoder and additional speakers needed for surround.

Your tastes in music and movies or skills in recording audio will determine what you choose. In order to listen to a hi-fi recording, you will need at least quality hi-fi stereo equipment. In order to enjoy the magic of surround sound, an A/V setup is needed. If mixing and recording is your forte, then tape decks, DATs or MiniDiscs are in your future. Let's take a brief look at each component:

Receiver: The audio market has all but migrated to the home theater arena since the spurt in surround sound. Most receivers on the market today are in fact audio/visual (A/V) receivers capable of enveloping us with one form of surround sound or another. More about receivers in Chapter 6.

CD Player, DVD Player or MiniDisc: Plastic plates full of digital data are in! This can be a simple one-disc unit, or a carousel system that can hold hundreds of CDs at once. The newest option is a digital video disc (DVD) player. It lets you play either a standard audio CD or a full-length movie with the same player. The MiniDisc is also gaining popularity since you can also use it to record onto discs.

LP: Yes, they're still out there. Prophetic spinheads are hoarding their vinyl treasures knowing the end is near. Records are more of an audiophile market, but some people argue it's the only playback medium that adds character to an album.

Tape Deck: Most component and rack stereo systems come with some type of cassette deck. It can be either an older single unit, or a dual unit for dubbing or to use as a carousel.

More about CD players, DVDs, MiniDiscs, LPs and tape decks in Chapter 7.

Surround Sound Decoders: These circuits are usually built into most A/V receivers. Dolby surround sound turns the encoded signals into something the rest of the surround equipment will comprehend. More on surround sound in Chapter 4.

Loudspeakers: There are many loudspeaker configurations these days. In the past, two large, towering monoliths that passed as speakers hovered in the living room. Companies still offer this option. However, with the multiple speakers needed for surround sound, these large beasts only add to the clutter. Manufacturers are popularizing a setup called *sub/sat* combos (subwoofer/satellite). This setup uses two to four smaller, inconspicuous speaker boxes placed about the room with one large, stealthy subwoofer cabinet disguised as furniture. More on this subject in Chapter 8.

Microphones: People do still use stereos to record music, events or even their own voices. In these cases, it is best to have a high-quality microphone.

Headphones: Headphones have a dual role. One is to keep the music all to yourself, and the other is to muffle irritating external noise such as cars, kids or people yelling for the music to be turned down.

Remote Control: Oh yes! How can I forget the remote control? Who wants to get unnecessary exercise each time someone yells, "Turn that down!"? Most A/V receivers typically come with a million-button remote to help you keep your hands in shape (at the very least).

REAL-WORLD STEREO AND SURROUND SOUND SYSTEMS

Hi-fi components are sold either as all-in-one packages, piecemeal, or as a system with separate pieces from the same manufacturer. These categories for stereos are *combination*, *separate component* and *rack* systems (component sets). Each type of set has pluses and weaknesses. Add to this surround sound A/V receiver combinations, and you have a confusing concoction of components. Let's try to clear up some of this audio consumer daze by looking at each type of stereo system.

Combination Hi-Fi System. Companies have begun to cook up a new smorgasbord of all-in-one stereo systems. These combination systems are the wrapped-in-a-single-box solution to simple audio equipment. They bundle together such components as a receiver (which includes an amplifier/preamp/tuner/decoder), tape deck (often dual), CD player, LP and speakers that are either integrated with the rest of the package or are detachable.

Pros: Combination systems are simple systems with straightforward controls. Setup is painless, as there are no connections to get tangled into. They usually come with a remote control that works every section of the unit.

Cons: If you want to customize, maximize or update, think again. This would mean purchasing an entirely new system. Also, the controls tend to break off.

Separate Component Hi-Fi System. Systems with separate components and enclosures are called (surprise!) *separate component hi-fi* systems. They can include (but are not limited to) a receiver, a tape deck, a CD player, a turntable, speakers and possibly a separate amplifier. Other components may include a digital audio tape deck, a Dolby Pro Logic receiver, a home theater surround sound system, DVD player, etc. Each component is then connected through a series of cables to synchronize your custom package.

Pros: The system is very modular in design. You can add components as they come out or as you can afford them. There is no limiting yourself to one company's products. A Marantz receiver can be set atop a Sony CD player, and JBL speakers can become your sound drivers.

Cons: Complex. Harder to hook up and control. More costly than a combination system.

Rack Stereo System or Component Set. Rack systems are extremely popular with beginner to mid-end audio consumers. This is a separate component system, but the same manufacturer makes each component.

Pros: Easy purchasing decisions (all their components or nothing). Warranties are usually better, and you can get any component fixed at one location.

Cons: Sometimes all of the components chosen by the manufacturer don't meld harmoniously. They may add a cheap set of speakers to a fair quality amp and a good CD player. If you want choice, forget it!

Home Theater Sound Systems. Home theater has created a new audio system category: surround sound! Instead of only two tracks and two speakers, surround records 3+ tracks and plays them back on 4+ speakers. The components needed for surround differ from regular hi-fi, but manufacturers are making receivers integrating the two into one handy package. You typically need a special receiver/decoder so the rest of your stereo understands the Dolby surround signals. Also, you will need four speakers or more (five is typical). Home theater is explained in detail in Chapter 4.

Pros: Sound signals from videos and TV are very dynamic and encompassing.

Cons: You have to purchase more equipment and speakers. If you have a combination hi-fi system, you will likely have to start from scratch.

WHICH SYSTEM SHOULD YOU PURCHASE?

Now the eternal questions:

"Regular Hi-Fi or Surround Sound?" This question has basically been answered for you already. Unless you buy a combination system that doesn't have surround sound, you will likely have an A/V receiver. Most receivers on the market are A/V and surround sound compatible. All you have to do is add the extra speakers.

"Should I Choose a Combo, Separate or Rack Hi-Fi?" You most likely have already decided this, but here is more data to help you. Remember these keywords for each:

> Combination System = Simple and Cheap
> Separate System = Choice and Expandable
> Rack System = Economical and Expandable

Now ask yourself, "Which one fits my needs?"

WHAT CONSTITUTES AN AVERAGE AUDIO SYSTEM?

Refer to Figure 1-2. Let's start with the taskmaster of your stereo: the receiver. It is a combination of a tuner, preamplifier and power amplifier. Because home theater surround sound is so popular these days, people are opting for a Dolby Pro Logic or Dolby Digital receiver/decoder instead of an old stereo receiver. It is actually difficult to find a two-channel-only receiver these days.

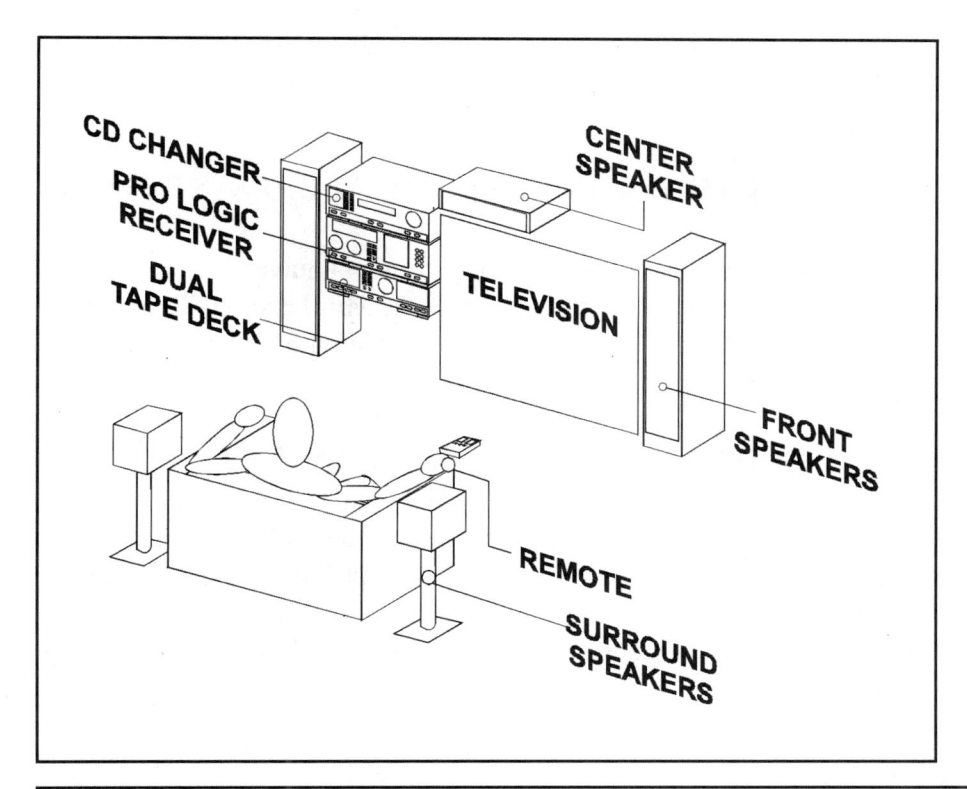

Figure 1-2. Average audio system.

Chapter 1: Introduction to Audio Equipment

Most people prefer the new CD format for listening to prerecorded music. CDs are readily available, aren't easily damaged, and they last. A typical audio system comes with a CD changer that can hold at least 5 CDs, hopefully more.

Recording (tape) decks are still popular as a cheap way to record or mix audio. A dual-deck unit is typical, with Dolby Noise Reduction circuitry.

Speakers are the components that require the most research. A poor choice of speakers can utterly ruin an otherwise perfect hi-fi system. Always go for a set of speakers that matches your amp. Plus, make sure you actually listen to them before buying. Try to be blind to the brand name and open to the sweet sounds that they produce. A good speaker package is ONE THAT SOUNDS GOOD TO YOU! Otherwise, the only choice to make in speakers is between surround and regular twins.

Whichever system you choose (combination, rack or separate), you need to be able to control the gadgets. Make sure all knobs, dials, etc., on the components are easy to use and interpret, and don't come off easily. Make sure you get a remote control that handles each component in the deal.

WHAT CONSTITUTES AN EXPANDED AUDIO SYSTEM?

If you chose a rack or separate audio system, you have room to expand. Maybe you want to add a subwoofer, DAT, MiniDisc, DVD player or LP. See Figure 1-3. If dressing up the system with beautiful, glowing glass tubes is your niche, then go for it.

WHAT CONSTITUTES A LIGHT AUDIO SYSTEM?

Combination hi-fi systems are a great deal. These elementary systems are aimed at the low-end market. They usually have simple controls and attachable or detachable mini-speakers. They are not appropriate for A/V applications despite what the ads say. So beware; they are not upgradable.

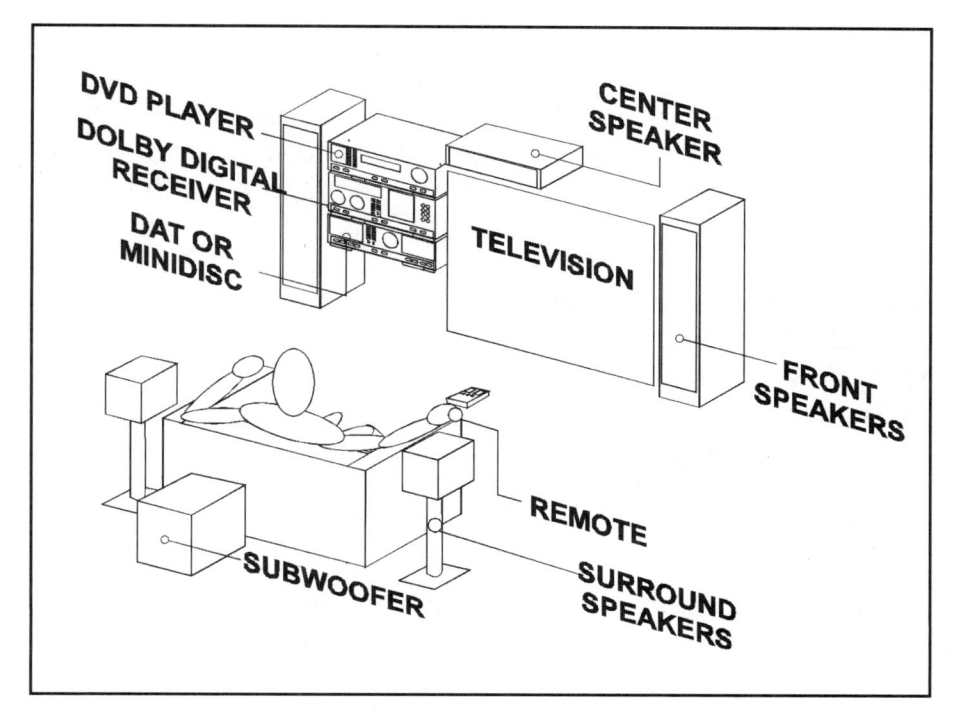

Figure 1-3. Expanded audio system.

Try to find a unit that contains all the features you want, including a remote control. Make sure to listen carefully to several units before making a decision. Once the purchase is made, you are stuck with the merchandise.

HOW YOU CAN IMPROVE YOUR AUDIO SYSTEM

There are many ways to improve sound quality. Some methods are simple and cheap, while others are exotic and expensive. Whatever the solution, your goal is to reduce distortion over every frequency band.

Easy/Cheapest: The simplest improvement that is the lightest on the budget is also the most obvious: SET UP THE SYSTEM CORRECTLY! This cannot be overstressed enough. This includes using the proper geometry for speakers, obeying the polarity of the speakers, making sure connections are tight and sure, matching impedances between the amplifier and loudspeaker drivers, setting the controls realistically (don't overdrive the bass!), etc. Follow this rule, (yes, I'm going to say it again): SET UP THE SYSTEM CORRECTLY!

Midrange: Follow an upgrade plan that you have laid out in advance. This may include purchasing an A/V receiver with Dolby Pro Logic, a more robust amplifier, a multidisc CD or DVD player, a slamming subwoofer, higher quality speakers, etc. Buy one component at a time, starting with a quality A/V receiver, then quality speakers. Move up to others as the budget allows.

Hard/Most Expensive: Use high-quality gold connections, Dolby Digital/THX components, DVD, DAT, MiniDisc, etc. Integrate the components into your surroundings or environment. This may involve building a home theater around the audio and video equipment. Who wouldn't want a soundproof A/V room?

25 AUDIO TERMS YOU SHOULD KNOW

Let's take a look at some terminology that you should know before you go shopping:

5.1 Channels: Dolby Digital uses five discrete channels as well as a dedicated subwoofer signal (that's the ".1").

Acoustics: The science of the production, transmission, reception and effects of sound. A more formal definition is, "The characteristics that control the reflection and absorption of sound waves in a listening area."

AM/FM: *Amplitude modulation* (AM) is the means in which an electrical signal is sent using the amplitude of the wave. *Frequency modulation* (FM) codes the information into varying frequency bands. The modulated electromagnetic waves containing the audio or video information are then sent to our antennas and tuners, which demodulate the signal back to sound or pictures.

Amplitude: The height of a wave. The higher the wave, the larger the amplitude. To amplify usually means to add to the amplitude of a wave or the level of power.

Analog/Digital (A/D): Analog is a continuously varying signal. It may be a constantly changing voltage, resistance or amperage. When we refer to analog in audio, we are pointing out the fact that traditional electronic components are producing constantly varying voltage levels. Digital, on the other hand, refers to a state that is either ON or OFF. There are no varying levels. However, by using several on- and off-state components, we can simulate a varying signal. In audio, "digital" means that the device uses digital electronics such as microprocessors and other transistorized two-state components. Digital is a by far better method of processing audio, as it does not introduce heavy noise into a system as some analog components do. It is easily compressed, which means more information can be squeezed into one spot. Most audio equipment is now shifting to digital electronics.

Audio/Visual (A/V) or Video: A/V refers to components that are used in both audio and video applications, such as home theater receivers.

Channels: A channel is a single path for electricity to travel. It is an independent-source amplification and output path inside your stereo. For example, the left channel contains the left microphone that recorded the original sound signal, the left recording, the left amplification, and the left speaker.

DAT (Digital Audio Tape): This is a special audiotape that records digitally. It stores much more information in a given size than a regular cassette tape. It can be used to duplicate exactly a CD or reproduce sounds with absolute crystal clarity. The popularity of this consumer audio fax machine has not caught on in the USA as of yet.

Digital Audio Broadcast (DAB): DAB is the future of radio broadcasting. This will be the radio equivalent of digital television. All audio signals will eventually be shifted over to digital. This means you will inevitably need to replace all of your tuners if you want to listen to radio stations.

Digital Video Disc or Digital Versatile Disc (DVD): This is the new compact disc on steroids. It is a CD the same size as an audio CD, except that it fits a whole high-resolution movie with Dolby Digital sound signals on one disc. You need a special DVD player in order to play the DVD discs. This is the future of A/V players.

Distortion: Any element that alters the faithful reproduction of the original sound. This is the mud that dirties up a shiny recording. It comes from either the original recording equipment or malfunctions and incorrect setups in your equipment. Annoying!

Dolby Surround Sound, Dolby Pro Logic and Dolby Digital: This refers to Dolby Laboratory's surround sound technology. Dolby does not make the equipment itself; the manufacturer has licensed Dolby's decoding scheme that produces encompassing sound. Surround sound typically uses 3 channels and 4 speakers; Pro Logic typically uses 4 channels and 5 speakers; and Dolby Digital uses 5.1 channels and 5-6 speakers. More in Chapter 4.

Driver: Drivers are the cones and domes inside the loudspeaker that actually produce sound. They use electrodynamics principles to push a surface back and forth, displacing air. Most loudspeakers have three or more drivers screwed into their surface: a woofer, midrange and tweeter. Each delivers its own frequency range.

Frequency: A wave is measured in cycles-per-second. This is its frequency. If an audio system's speaker kicks back and forth 500 times a second, then its frequency is 500 cycles/sec, or 500 Hertz (Hz).

Home Theater: This is a new category of consumer electronics that has boomed over the past decade. It includes a large television or monitor for picture and a sound package that is typically surround sound. The idea is to emulate a real theater in your living room.

MiniDisc: This is a record-and-play CD-type disc. It uses a small 2½" disc contained in a cartridge that is placed in a home deck, portable player or car audio system. It will record or play up to 74 minutes of compressed audio.

Receiver: This stereo component typically contains the tuner, preamplifier, power amplifier, controls, decoders and signal processors. It is the heart and soul of your audio system.

Signal-to-Noise Ratio (S/N or SNR):   This is simply the ratio of desired output signal to undesired noise created by the system itself.

Sound: Sound is energy transmitted through the medium of air. We sense this energy by the varying pressure waves hitting our eardrums. The energy is then converted to an electrical signal our brain interprets as sound information.

Sound Stage: With stereophonics, we are recreating a three-dimensional space in which the musicians and singers are "located." By placing two or more microphones upon a stage and recording two different tracks of sound, we are able to recreate that sonic space. This includes capturing subtle nuances such as the acoustics of the room. When we play back these two tracks on two appropriately-placed loudspeakers, we are able to recreate the effects of the sound stage right in front of us.

Stereo: *Stereo* can refer to stereophonic technology or to the actual stereophonic equipment. It indicates a multichannel system, usually with two channels. This recreates the sound stage of the original recording and adds dimension and texture to otherwise mono-sounding music or vocals.

Subwoofer: This is the driver that recreates the lowest frequencies in an audio system. It is usually roosted in a separate, "end table-like" cabinet. When you hear (and feel) that THUMP, THUMP in recordings being played on a subwoofer-equipped system, it would be from this driver.

Transducer: A transducer is a device that changes energy from one form into another. Microphones transform acoustic energy into electronic energy and speakers conversely change it back.

Tuner: A tuner is the part of a receiver that locks onto a channel or station and demodulates its signal. Then the rest of the stereo can use the sound information that was extracted from the radio station's signal.

Watts/Ohms: Wattage is the measurement of power. Typically, an amplifier is rated in wattage, showing the power that it is able to sustain when sending a signal to the speakers. Ohms are a measurement of resistance or impedance. Speakers are typically rated as 4, 6 or 8 Ohms.

Chapter 1: Introduction to Audio Equipment

WRAP UP

Now that you have finished this chapter, try to decide if you should build a component system brick by brick or buy a rack system to save a few hundred bucks. If you're a non-audiophile, a combination system may be your cheapest, hassle-free option. If you want a quality system for hi-fi and home theater applications, and have this book down pat, be a brave, sound soul and build it one component at a time, starting with a high-quality A/V receiver. There's always the next paycheck to blow on more components.

A Brief History of Audio Equipment

It is hard to imagine that TV was once unavailable. The only form of entertainment you could bring into your home was radio. A centralized radio station would use the airwaves to send AM signals to our homes. The signal would contain the sounds of that age, including news, songs, talk shows, weather reports and theatrical radio dramas. The entire family would huddle around huge wooden cabinets filled with tubes and dreams.

The quality of the systems increased as electronics advanced. High-fidelity (hi-fi) stereo equipment was eventually discovered. People didn't think they needed a higher-quality unit until they experienced its spacious ambiance and accuracy. Hi-fi soon became all the rage. Up until the last decade or so, it remained a permanent fad. Then the market began to slump along with the drowning economy.

In the 1980s, a move toward home theater started. This proved to be the saving grace of the audio industry. Audio hi-fi equipment was married to video equipment just like the mega-amalgamations of companies in the '80s. This saved the industry as people were now able to buy a surround sound system that would stun anyone with its clarity, realism and fidelity. The audio industry was reborn.

CHAPTER 2
SOUND REVEALED

"All sounds have been as music to my listening ..." Wilfred Owen

What is sound? It is vibrations transmitted through the air. Sound is the energy that emanates from an object moving back and forth. See Figure 2-1. Let's use a stick striking a snare drum as an example. When the stick contacts the drum's skin, energy is transferred to the surface, bowing it inward. When the stick lifts off, the skin returns to its original position; but it now has the stick's energy added to it. Air molecules are pushed along with it. The drum surface, depending on its elasticity, springs back and forth, and back and forth, each time displacing air molecules. This creates high-pressure areas and low-pressure areas.

The energy from the drum skin is then sent out to the surrounding air as sound waves. These waves move outward in all directions, eventually

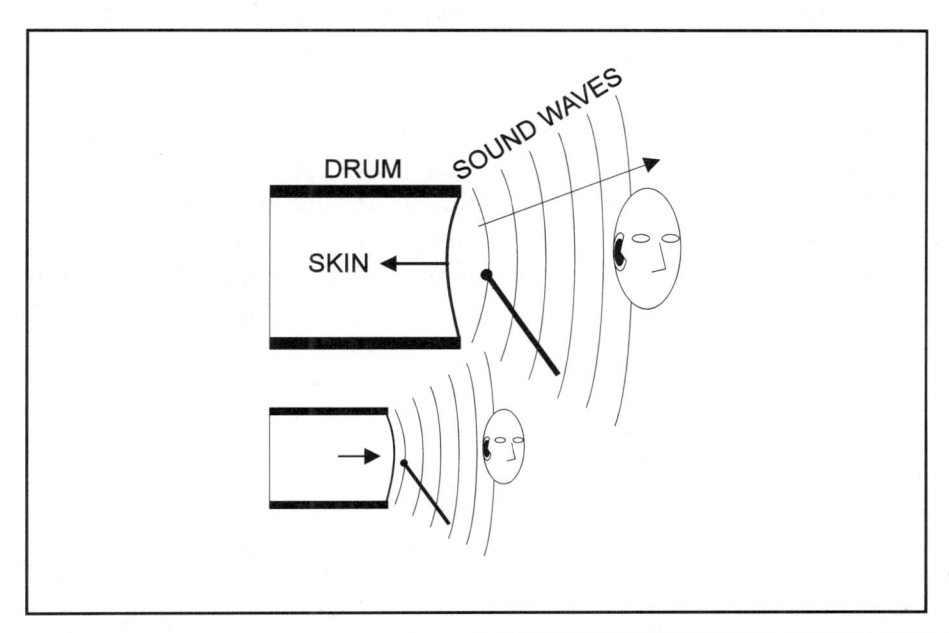

Figure 2-1.
What is sound?

NOTE: Despite the thunderous and screeching sounds of space ships in the movies (*2001: A Space Odyssey* being an exception), there cannot be sound in space. Space is a vacuum and void of air or molecules (matter). Air, or other forms of matter, is needed to transfer the energy of sound. Air is called a medium, and energy is transferred through a medium.

reaching your ear. So, when you hear a sound, you are actually sensing the energy (in the form of pressure and rarefaction waves) being sent to your ear from a vibrating object.

MEDIUMS, SOURCE POINTS & RECEIPT POINTS

Sound is energy. It has a source point where the energy is first created. In an ocean wave, this source energy would come from the wind. The kinetic energy then needs a medium to travel through; in most cases, this is air. In the ocean, the medium would be water. The medium moves the kinetic energy along by bumping one molecule into the next. Once the energy travels a distance, it will undoubtedly come across a receipt point. In sound, the receipt point is your ear drum, which converts the kinetic energy into electrical energy that your brain will understand.

WAVES

To understand amplitude and frequency, let's best look at the principles of energy transfer through waves. More specifically, we are addressing sound waves. Don't be scared. A physics degree is not necessary to understand the sound coming from your hi-fi. Basics are basics, and your audio components are all built upon the same acoustic and energy ABCs.

A wave transfers energy through a medium. The medium is used to bump molecules back and forth to carry the energy seemingly outward or forward. When we are dealing with oceans and waves of water, the medium is water. Sound waves are similar to ocean waves except that they drive the energy forward through the medium of air. They are both using the medium to deliver energy to a distant location.

SOUND WAVES

A sound wave is built of two pressure zones. One is high-pressure compression of the surrounding air, and the other is a low-pressure rarefaction section. Rarefaction and compression make up the complete sound waves. See Figure 2-2. The back-and-forth physical movement

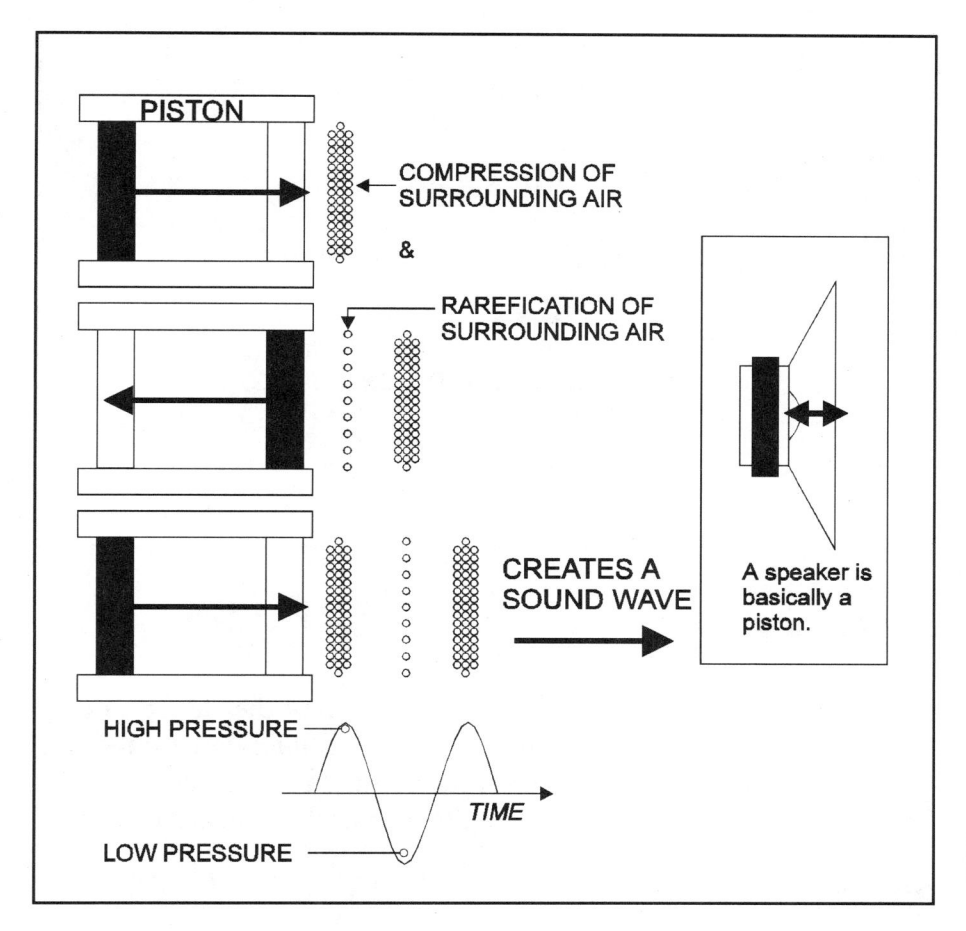

Figure 2-2. Sounds waves.

of an object can create positive and negative air pressures. In the case of a stereo, the driver moves a cone back and forth, thrusting and yanking the surrounding air. This sends out the sound waves we hear.

FREQUENCY & PITCH

An object or wave that is moving back and forth or pulsing (cycling) is measured in cycles per second, or Hertz (Hz). This is the object or wave's *frequency*. See Figure 2-3. A frequency of one cycle per second is one Hertz (1 Hz). With sound, we are dealing with frequencies between 10 Hz up to about 20,000 Hz (20 kHz). This is the limit of human hearing. Frequency and pitch are often used interchangeably with sound.

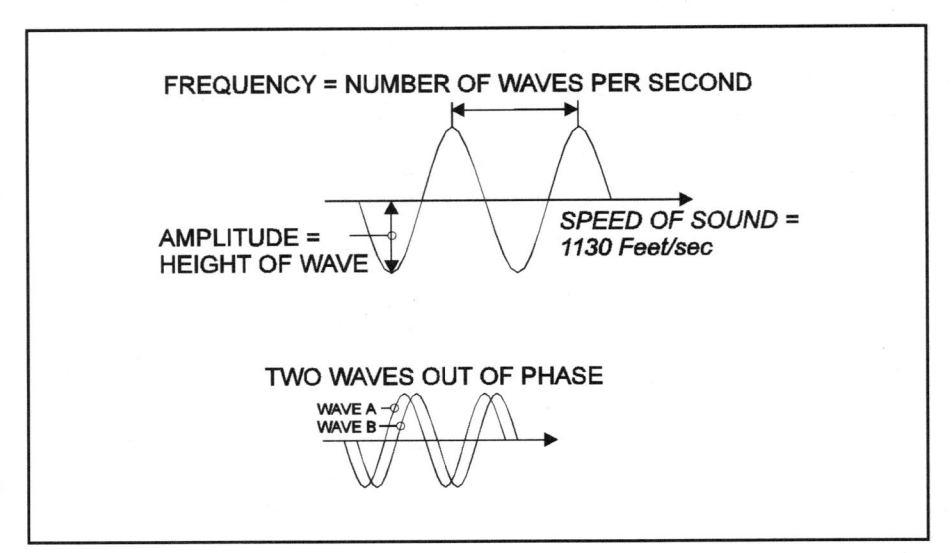

Figure 2-3. Measurement of sound waves.

AMPLITUDE

The distance an object is moved back and forth or the strength of the pulsing is its amplitude. The higher the amplitude, the stronger the wave coming at you; thus the louder the noise. In fact, an audio amplifier increases the sound signal's amplitude in order to raise the volume.

PHASE

Two waves can be exactly alike yet out of phase. This means that one of the waves is slightly behind the other. If these two waves mix, there will be constructive and destructive interference. Phase is the result of a slight time delay. In the next chapter, you will see how this is used to help create a stereo effect.

SPEED OF SOUND

A sound wave travels at approximately 1130 feet per second. Watching a lightning storm or explosion from a distance will demonstrate the sluggish pace of sound compared to light's 186,000 miles per second. When dealing with audio equipment and stereo mechanics, we are dealing with relatively short traveling distances (a mere living room). But it is important to understand that merely moving a speaker a few feet off-mark will cause the waves to reach you that much faster. This will discolor an otherwise perfect stereo effect.

Waves are measured in three different ways. One way is by wavelength (λ). This is the distance between the tops of the waves. Next is by the speed of the wave. Last is frequency. Here are the calculations used to determine each:

$$\lambda = c/f$$
$$c = f/\lambda$$
$$f = \lambda/c$$

Where λ = wavelength, c = speed of sound in feet/sec or meters/sec, and f = frequency.

So, if you want to figure out a sound wave's wavelength, use the first formula. Example: a 3,000 Hz tone traveling at 1130 ft./sec. would have a wavelength of:

$$\lambda = 1130 \text{ divided by } 3000 = .37 \text{ foot}$$

NOTE: Sound moves roughly 1 foot in one thousandth of a second. At 20 Hz, there is a new wave cresting every 65 feet. But at 20,000 cycles per second, the top of a wave hits every half an inch.

DECIBELS

Sound pressure level (also called loudness) is measured in decibels (dBs). A decibel is a unit used to measure relative changes in loudness. The human ear can perceive a difference of 1 dB. Our ear reacts logarithmically to sound. This means that when the sound pressure or power is doubled, we do not hear it as twice the volume, but rather one tenth louder.

Decibels are not a set measurement like an inch, but rather are the logarithmic difference between any two measurements. In measuring sound pressure level, we use a reference level of zero (the threshold of human hearing) to come up with a relative list of various sound levels:

Explosions	140 dB
Threshold of Pain	130 dB

CAUTION: The sound pressure level is not the only factor used in determining if a sound is damaging to your ears. A 1000-Hz tone at 120 dB will make you deaf. However a 20-Hz tone at 120 dB is normal. A train pulling into a station or a bus pulling up to a stop hits these high levels easily, though briefly.

Airplane at 18 Feet	120 dB
	110 dB
Thunder	100 dB
Train Whistle at 500 Feet	90 dB
	80 dB
Loud Music	70 dB
Normal Speech at 3 Feet, or Soft Music	60 dB
	50 dB
	40 dB
Very Soft Music	30 dB
	20 dB
Leaves Rustling	10 dB
Threshold of Human Hearing	0 dB

Remember, the power level needed to create a sound that induces pain is actually 1,000,000,000,000 times that of the threshold of hearing.

WHAT A HUMAN EAR CAN HEAR

Human ears have a very limited hearing range compared to some animals. The human ear's range is from 10 Hz to 20 kHz. We hear best in the 1000 Hz-to-5000 Hz range. A good high-fidelity audio system will reproduce these levels with little distortion.

The loudness of a sound determines the ability to hear, as well. See the previous section for a rundown of sound pressure levels measure in decibels.

NOTE: Human beings don't always use just their ears to pick up sound. Very low frequencies are actually perceived with our faces and bodies.

When we are talking about hi-fi equipment, we are dealing with levels of 10 to 20,000 Hz reproduced for our listening pleasure. Loudspeakers actually use several different drivers to reproduce these various frequency ranges. There is a subwoofer that booms out the low 10-to-100 Hz frequencies, then the woofers which deliver between 20 and 2000 Hz. What we perceive as voice is generally vibrated through the midrange drivers (between 800 and 10,000 Hz). Then there is the high-end treble driver, also known as a tweeter (between 4,000 Hz and 20,000 Hz). So, when we are listening to music, we are actually listening to a combination of drivers thrusting their multifrequency waves at us.

> **The Gift of Superhuman Hearing**
>
> Some people have extended-frequency hearing, or are able to discern very specific sounds. Being able to pick a single frequency out of a jumble is both a gift and curse. It is a gift for hearing far-off sounds or sounds caught in a mixture of background noise. It is a curse for enjoying a quality recording on a cheap stereo system. So, if you have this type of discerning ear, make sure you listen carefully to equipment before making purchases. Expensive, high-quality equipment may be the only way to sate those superhuman sound tastes.

ACOUSTICS

Acoustics are the science of sound. It is the study of its production, transmission, reception and effect on people. Acoustics are also a means of controlling the characteristics of sound. This is accomplished through "listening room" design. This includes harnessing the sound wave reflections and dampening unnecessary sound elements through absorption. Quite a task if you think about it. In this book, we are dealing with the acoustics of someone's personal listening room, as well as the acoustics that are subtly sculpted into the sound recordings.

HOW SOUND IS CREATED IN AUDIO EQUIPMENT

Today's electronic audio systems use basic sound principles just like our drum example in the beginning of this chapter. There is one difference: with the drum, we are changing mechanical energy (the hand swinging the stick to hit the drum) into acoustic energy (the air molecules vibrating).

With electronic systems, we are changing acoustic energy into electrical energy, storing or amplifying it, then changing it back to acoustical energy with the use of electromagnetic devices (speakers). The audio equipment changes the energy form to suit the electronics. See Figure 2-4.

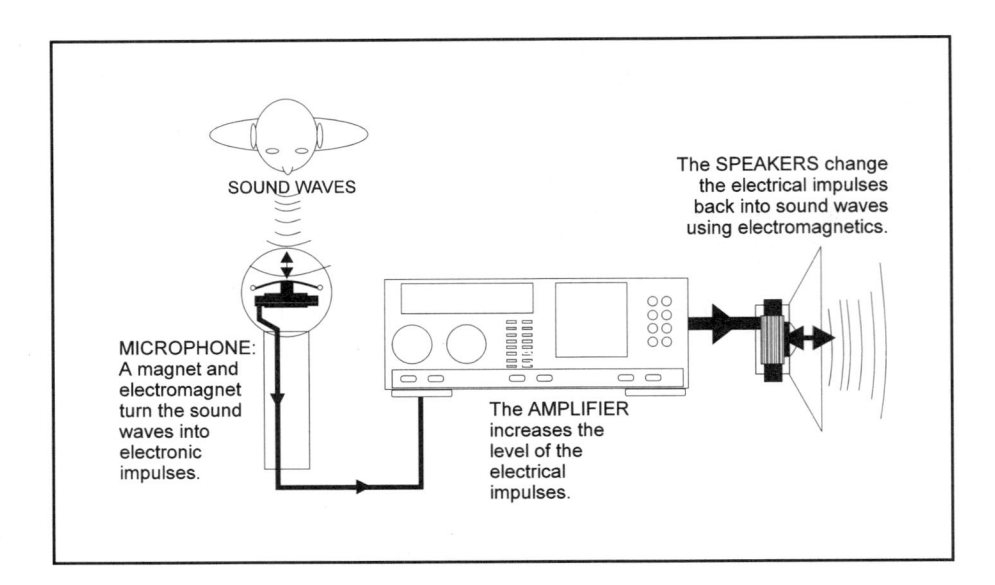

Figure 2-4. How audio equipment makes sound.

ACOUSTIC PROPERTIES IN AUDIO SYSTEMS

Audiophiles pride themselves with having their own audio nomenclature to help describe the acoustical quirks and qualities of audio equipment. It may seem like Greek to the non-initiated, but once you understand, you will be able to communicate "stereo-ese" to salespeople and friends much more easily. Here is a cross-section of this "sound" language:

Acoustics: When we are referring to *audio system acoustics*, we are talking about the properties or character that the sound takes on in a room. Does it sound like you are in an open space or closed in? Can you hear the reflections from obstacles in a room?

Airy: Spacious, open feeling to the sound. This means there is a very accurate reproduction of high frequencies.

Bassy: The low frequencies are too heavy or overpowering.

Blurred: This means the stereo imaging is not set properly. Unfocused.

Crisp: Extended high frequencies.

Detailed: This is a system that does a good job of reproducing minor details in the music. A guitar is a guitar, a drum is a drum, and you can tell whether a singer is good or bad.

Frequency Response: This refers to how well the device is transmitting different frequencies applied to it. If your speakers have good frequency response, they are reproducing various frequencies without getting bogged down.

Hot: Means overly high and overpowering in a specific frequency range. A hot tweeter would be screechy and a hot bass would be bumping you out of your seat.

Loudness: This is a measurement of how loud a system is, usually in decibels. It is the amplitude of the sound wave or the intensity of the energy reaching your ear.

Sonic Impact: The impact the sound has on your emotions. Rock creates a different sonic impact than classical music.

Timber: This is the tonal musical quality that enables a listener to distinguish one instrument from another. An example is distinguishing a violin from a flute when they are playing the same note.

Transients: A quick impulse or disturbance in a system.

Treble: The high frequencies in a playback, often reproduced by your tweeter and midrange.

HOW SOUND IS RECORDED

Microphones are transducers. A transducer is a device that transforms one form of energy into another. The mike transmutes the sound energy into electrical energy through the use of electrodynamics, just like your speakers. The electrical energy is then sent to a preamplifier and either recorded or routed to an amplifier and speakers.

HOW SOUND IS STORED

The storage of sound is the largest landmark of audio equipment discoveries. Recording devices take the electrical impulses created from by microphones and store them magnetically onto a special metallic-coated tape, in grooves on an LP, or optically as pits and lands on a compact disc. When you are ready to listen to the prerecorded sounds, the electrical impulses are converted back into electrical sound impulses and sent to an amplifier and speakers. See Chapter 7 for more information on sound storage.

HOW SOUND IS PLAYED

Speakers are also transducers. They change electrical impulses back into acoustical energy so our ears can understand the output. A signal source is fed to an amplifier that takes a weak signal and increases it. The energy is then used to energize an electromagnet (mounted inside a permanent magnet), which in turn drives a cone back and forth to create the pressure and rarefaction. Thump! Thump! That's what we hear.

SIGNAL-TO-NOISE RATIO

So far, there is no such thing as a perfect signal. There is always some amount of noise along with the original vibrations. The ratio of actual signal to noise, or distortion, is called the signal-to-noise ratio, or S/N. It is usually measured in decibels. A 16-bit CD player, which reproduces a very clean signal, inherently has about 96 dB S/N. It is important to look at this specification on your amplifier and other equipment to get the best-sounding stereo possible.

WRAP UP

The science of sound is not all that difficult to understand. It's like being back in school again, isn't it? But at least you can put this knowledge to good use in your audio system, especially when picking out your system.

Why We Convert Sound into Electronic Signals

Sound waves weaken exponentially with distance. The farther you get from the source point, the lower the volume; until eventually, even a yell becomes a whisper. Electronics solves this distance problem. It is easier to convert sound waves into electronic waves and send them to a remote area than it is to build a city-size speaker. By using electrons instead of sound pressure waves, sound can travel an indefinite distance. One problem: you need expensive equipment to convert the electrons back into sound waves.

Another reason for this conversion is enable us to store sounds. By turning sounds into electronic waves, we can utilize modern media to save the sounds.

CHAPTER 3
STEREOPHONICS REVEALED

"Stereo does not equal monophonic times two." Randall Haffner

What is the definition of stereophonics? It is the study of multichannel, 3D sound and the electronic audio equipment used to reproduce it. Stereo hi-fi equipment is designed to evoke a sense of separation, like that of the original recording. In other words, it is a way to recreate a sound stage, including the ability of the listener to determine the original source location of specific sounds. See Figure 3-1.

Why should you explore the world of stereophonics and its workings? Because it is important to master the science of stereo in order to recreate the desired sound stage that the recording artist intended you to hear.

NOTE: The derivation of the term "stereophonics" is Greek: *stereo* is "3D" and *phonic* is "the science of sound." Therefore, *stereophonics* is the science of 3D sound.

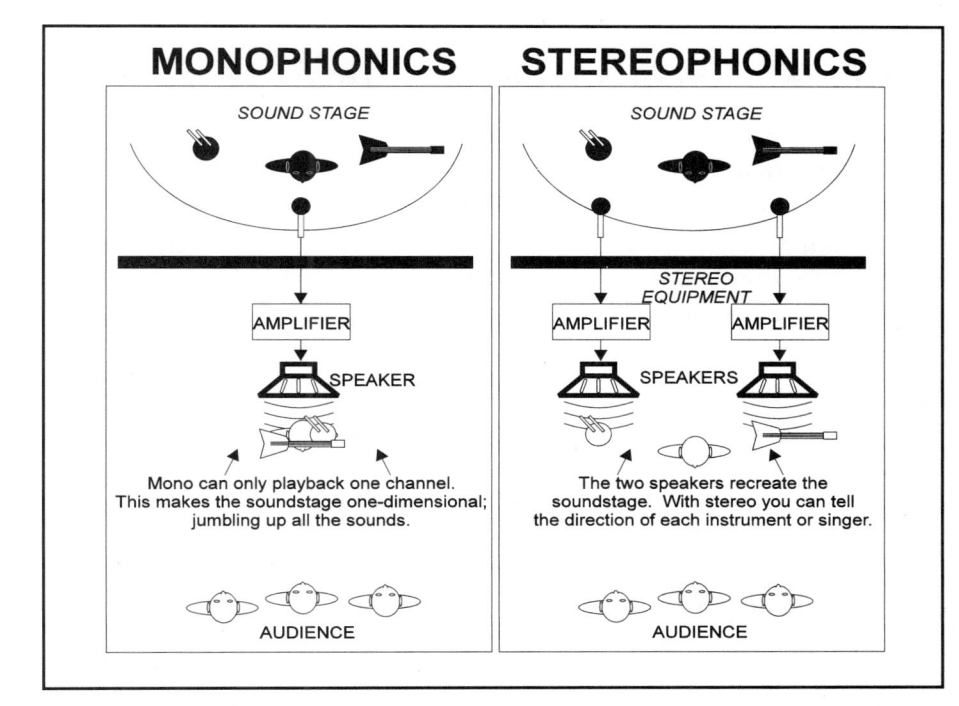

Figure 3-1. Monophonics and stereophonics.

Chapter 3: Stereophonics Revealed

Simply knowing the correct position for a speaker can save you thousands of dollars: money that could have been spent trying to improve your system with costly upgrades. You will even be able to figure out the complicated surround sound system and learn how to cram an entire concert hall into your living room. On with the expedition.

CHANNELS

A channel is a single path for transmitting electrical signals. If you have a one-channel audio system, there is one source (microphone), one recording channel (cassette or other), one amplifier channel, and one speaker for playback. (Note: one channel can be sent to several speakers, but they all play the same channel.)

Multichanneling makes several recordings of the same sounds. It's like having two or more tape recorders running simultaneously. However, because the channels are recorded with more than one microphone (placed apart), the channels are not exact duplicates of each other. There are subtle phase, acoustic and pressure level differences. You will see why we use multiple channels in a moment. For now, remember that each channel is a separate recording or path.

LIFE BEFORE STEREO: MONOPHONIC

Before the dynamic technology of stereophonics, there were *monophonic* and *monaural*. "Monophonic" means "one sound" or "one ear." To further clarify this, see Figure 3-2. Mono systems utilized only one channel of recording and one channel of playback through one or multiple speakers. It sounds like the speakers are in another room and you are listening through a small hole in the wall. This was just fine for most people until they heard a quality stereo hi-fi system. Once stereo was the standard, listeners were treated to a dynamic sound stages right in their homes.

HINT: Some low-cost VCRs and audio equipment still use monophonic. Why? It's a cheap sound solution. Do your ears a favor and pay the extra $40 for a stereo hi-fi VCR.

THE BASICS OF STEREOPHONICS

Mono is flat and one-dimensional, so engineers came up with a very elegant solution to add spatial texture and reverberation to soundtracks:

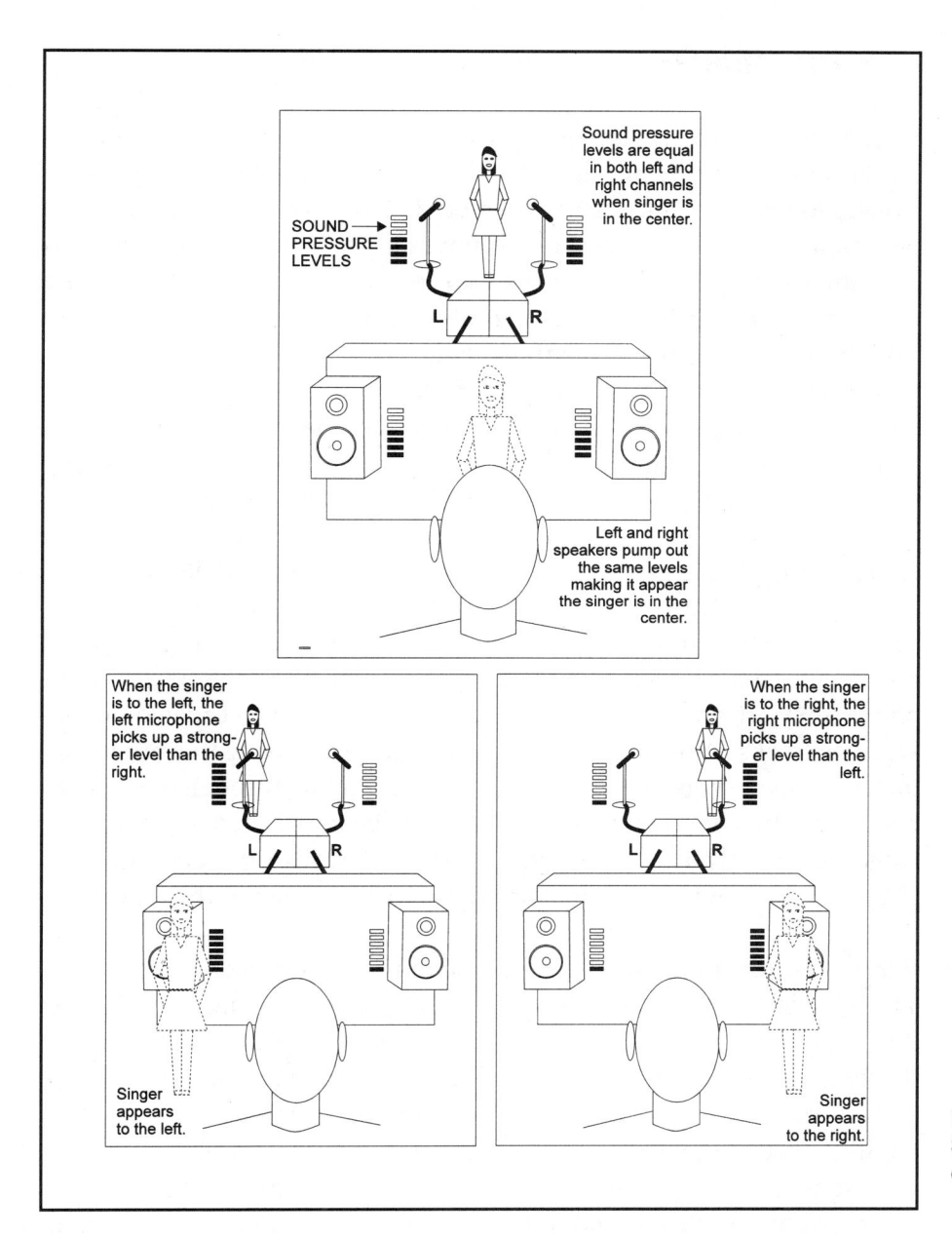

Figure 3-2. Stereo intensity explained.

they added another channel to the mix. *Stereo* reproduces the sensation that you are listening to a specific instrument/voice or combination in a specific space (room), with all of its acoustic properties simulated. But how does it work? How can there be such a dynamic difference by adding another channel and speaker?

Chapter 3: Stereophonics Revealed

HOW THE MAGIC IS PERFORMED

Stereo theoretically separates each sound source and places it in three dimensions. For example, a guitar could be on the left section of stage, a piano to the right, and a singer centered. How does stereo distinguish direction? Remember that stereo is not mono times two. There are two recording channels. This means two microphones are recording the sound pressures. They are placed at a distance from each other and usually reside in front of instruments or voices being recorded. Now let's look at an example. Refer to Figure 3-2. Imagine that a set of microphones is placed on a stage. Each microphone is recording one track; like having a tape made of each microphone's input, except they are on the same audio tape. This is called *intensity stereo*.

Refer to Figure 3-2. In the first example, a young female vocalist is singing center stage. The two microphones are equidistant from her. The sound level (measured in dBs) and frequencies coming from her mouth are sent to each microphone. When these reach each microphone, they are both at the same level. They are then recorded onto two separate tracks. Now, imagine hi-fi equipment reversing this process in the next room. You can hear the two tracks playing through two speakers (place apart and facing you). What you hear, or perceive, is that the singer is directly in front of you. Why? Because the equipment is playing back sound levels that are equal to each other at the outputs.

For another example, the singer is moved to the left side of the stage. The left microphone is now receiving a much stronger sound level (higher dBs) than the far right one. The left track will now record stronger decibels than the right. In the next room on the playback, the left speaker is mirroring the higher left levels and weaker right levels. This gives the illusion that the person is standing on the left side of the stage. The reverse would also be true: place her on the right, and we perceive her on the right. With a monophonic system, there would be no way to determine the location of our Diva on the stage.

ANOTHER WAY STEREO IS ACHIEVED

Law of the First Wave Front: "The first sound to arrive at the ears determines the perception of direction to the source."

The easiest way to explain this law is with an example. Refer to Figure 3-3. If there are two channels recorded when a piano is being played in the center, the instrument (upon playback) will sound as though it was placed in the center of the two stereo speakers. Now, if you apply the law of the first wave front, you can make it seem like the piano is coming from the left speaker, even though they are still the same signal. Delaying the right channel's wave by 5 milliseconds magically moves the piano to the left.

STEREO IMAGING

You now know the acoustic engineering tricks being played on your ears. But how can you hear a singer in front of you, a guitar to the right, and a piano to the left? Simple. Each sound carries its frequency, amplitude and decibel information to both microphones, as in the example

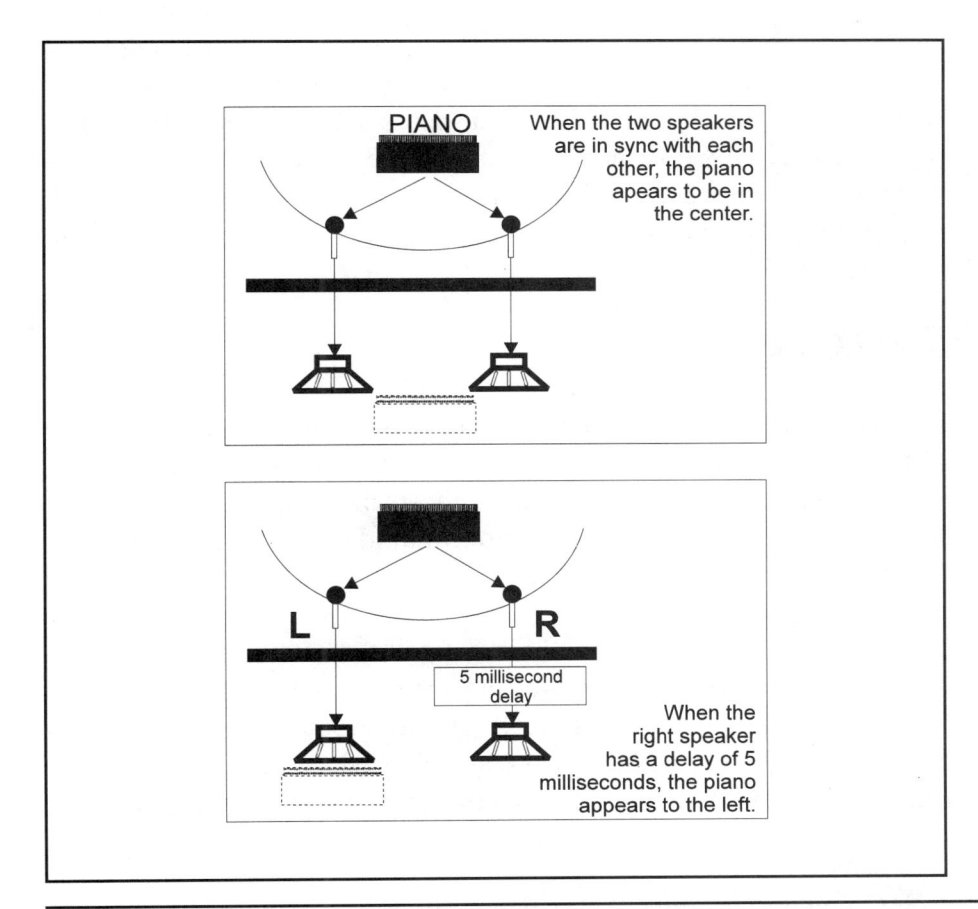

Figure 3-3. Law of the first wave front.

Chapter 3: Stereophonics Revealed

37

with the singer. The ingredients are already there to create an entire sound stage. The recorder merely has to make a copy of the left and right frequency/amplitude information. Then we set up our stereo systems to be conducive to the microphones. Hit *play*, and the entire sound stage is recreated with two tracks (channels) and two speakers. You know how a guitar sounds, so your mind chooses all the guitar frequency/amplitude information from each channel. The stereo effect does the rest. You can determine which instrument is in which position on the stage.

REVERBERATION MAKES THE SOUND REAL

Reverberation consists of the random sounds that surround the listener in a live concert hall experience. It is the hidden equation in stereo imaging. These are the sounds that are launched out to the audience, auditorium, concert hall or studio, and are bounced back to the microphones that are recording the two tracks. In doing this, the sounds inadvertently carry distance or spatial information to our brains upon playback on our hi-fis. We can *hear* open spaces or small obstructions simply by perceiving the sounds bouncing off the surfaces. Without reverberation, the recording and playback experience would appear flat, muddy, dry, plastic and dead. With reverb, the recording seems alive, *real* and full of depth.

STEREO GETS COMPLEX

Enhancing two-channel stereo technology is becoming more and more difficult. There is not much more that will polish up the way we listen to sound with two speakers. What engineers did next was to take sound to a new plateau. I am talking about multidimensional sound, commonly known as *surround sound*.

Surround sound puts you into the middle of the action instead of placing you in the "audience." Two-channel stereo cannot not fool a human being into recreating an accurate 3D image and placing themselves in the middle. For example, a two-channel speaker system can't create the illusion that every sound frequency is coming from behind or to the extreme sides of you. Some frequencies can emulate this, but not the entire band.

Surround sound took over all of the complex tasks required to place the listener in the sound field, much the same way fuel injection took over complex tasks that a carburetor couldn't handle any more. Surround sound sorts directional information and, using a computer chip, encodes and decodes it.

SURROUND SOUND

The goal of surround sound is a system that enhances the spatial imaging capabilities of the playback system, or to create a sound field that can surround the audience with direct sound as well as ambient sound from all directions. Dolby has developed several multichannel systems that meet this goal.

How does surround work? It uses 3 to 5 channels of audio and 4 to 6+ speakers that encircle your body at various locations. Regular surround sound uses 3 channels and 4 speakers. Pro Logic surround sound uses 4 channels and 5 speakers. The ultimate surround technology available right now is Dolby Digital, which uses 5.1 channels and 6+ speakers. See Figure 3-4 for a general explanation, and Chapter 4 for a more detailed look.

RECREATING EVENTS

Make sure you purchase a stereo that recreates the events you want to hear. Each recording has its intended effect upon the listener. Perhaps it places you in a certain sound stage or makes you listen at specific levels. Live recordings and music created in a special room or concert hall contain character because of the reverberations being laid into the recording. This adds to the ambiance and the sonic illusion, and makes you believe you are really in *that* space. Some recordings are designed for loudspeakers and others for headphones. A live recording of a concert on a stage is a great test of a stereo's capabilities. When a performer moves across the stage with an instrument strumming away, this effect is preserved for us to hear. It gives the recording animation, life

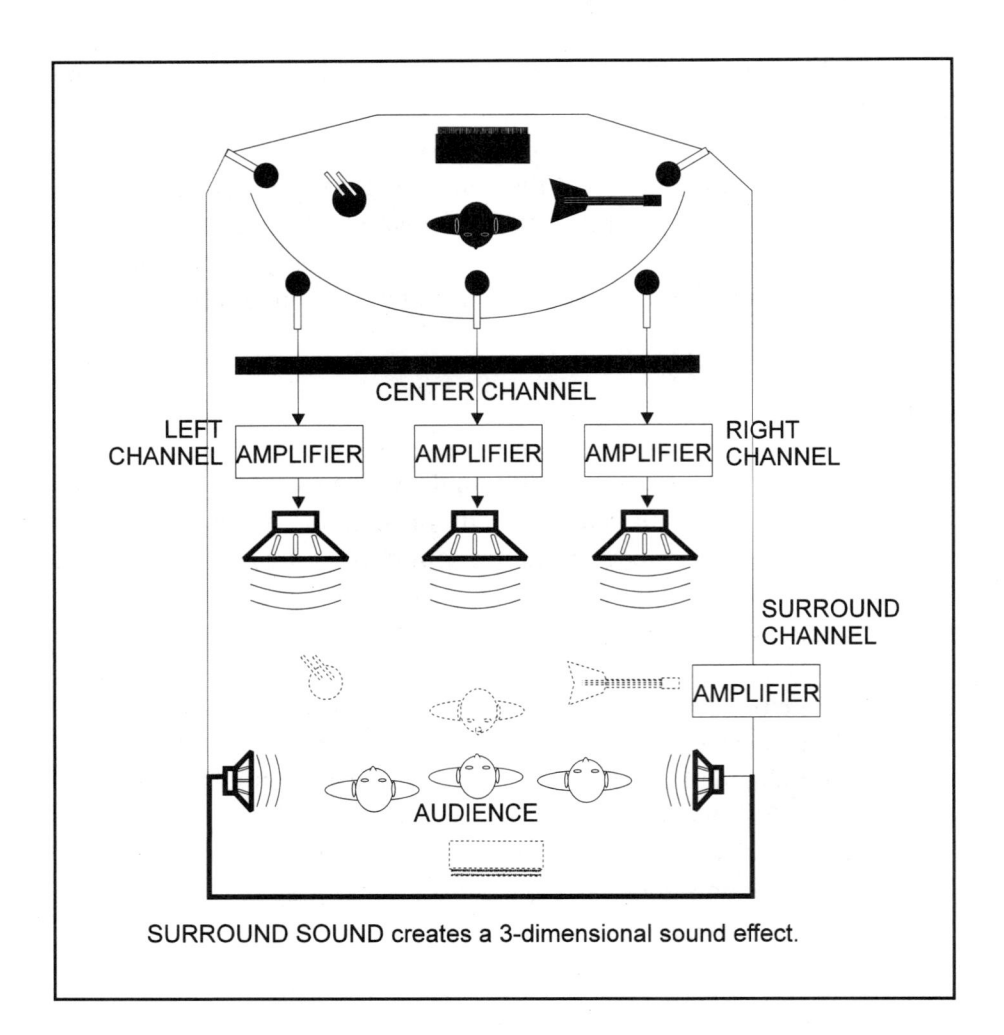

Figure 3-4.
Surround
sound
(4-channel).

SURROUND SOUND creates a 3-dimensional sound effect.

and spirit! When buying a stereo, make sure it actually recreates the event you want *in stereo*.

WRAP UP

As you can see, stereophonics is not a complex technology. Simply remember that stereo imaging uses multiple channels to record the original acoustics of the sound stage where the music or sounds were recorded in the first place. This allows for the playback of the spatial information with our hi-fi systems. Just like hosting a band in the living room, only cheaper!

CHAPTER 4
HOME THEATER SOUND

"The first thing in one's home is comfort; let beauty of detail be added if one has the means, the patience, the eye." George Gissing

How about a night out at the movies? No way! Why should you empty your pockets for two hours of being cramped in a room full of strangers? Bring the theater home.

The hottest area of home theater right now, and the one with the largest infusion of hi-tech gadgetry, is the "A" in A/V — namely, *audio*. TV monitors have yet to match the "movie experience" with a jumbo, crystal-clear video picture. Someday. However, most home theater surround sound systems rival or exceed today's cinematic audio equipment. Imagine a 10,000-square-foot movie theater squished into your living room. What a way to watch those favorite flicks! No icky and sticky seats, either.

THE SOUND EXPERIENCE

If you are a viewer that covets sound, can you imagine watching *Star Wars* without ever hearing a single laser-cannon blast? How about *Casablanca* with the absence of even one note of music? It would be sacrilegious, to say the least. A quality home theater sound system includes equipment that will deliver half the experience of the movie director's creation.

How does surround sound work? How can you drive speakers to rock the house with thunderous bass explosions? How do you make a sound field for immersing your mind in memorable movie scenes, even if only to hear a romantic secret whisper? What are Dolby surround sound and THX, anyway? On we go with home theater stereo.

SURROUND SOUND TAKES OVER

Because the popularity of A/V is so widespread, almost every audio product is now geared toward the better half of the A/V marriage. Surround sound is here to stay, so it's better to buy A/V equipment now instead of an old two-channel stereo. In fact, it is almost to the point now where it is difficult to buy regular hi-fi units except in the higher-end markets or from diehard manufacturers.

HOW SURROUND SOUND DIFFERS FROM NORMAL STEREO

Most stereos are two-channel. They emit their sound waves out of two loudspeakers located in front of the listener. The person must be in the stereo's sweet spot to receive the desired stereophonic effect. See Figure 4-1. This is between the two speakers and back several feet. If you are off to the side, well... next time call for your seats sooner, as your sound experience is now going to be lousy. Outside the sweet spot is a sort of sonically-shaded area, where the sound stage is dark to outside listeners.

Surround sound uses 3 to 6 channels to create an expanded sound stage. This creates a much larger sweet spot, so your entire mini-theater audience can enjoy the full effect of the soundtrack.

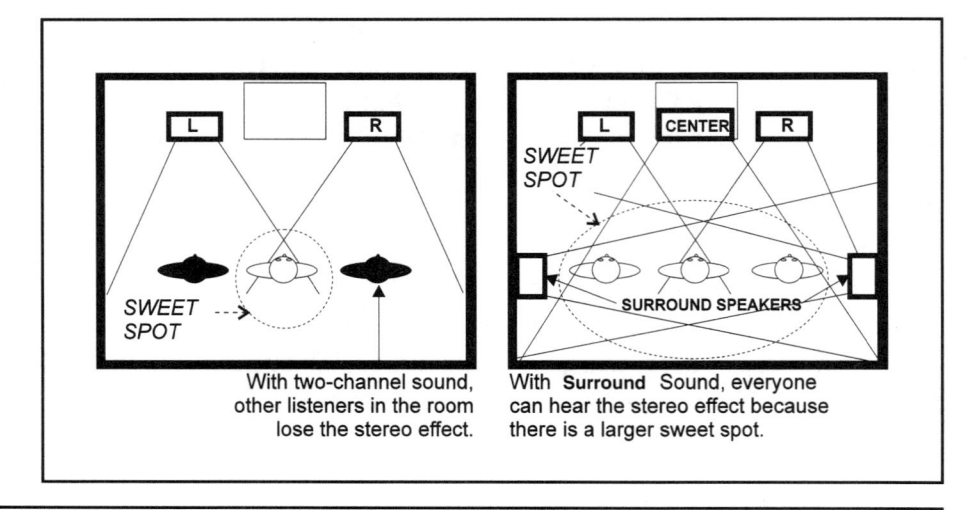

Figure 4-1. Sweet spots.

With two-channel sound, other listeners in the room lose the stereo effect.

With **Surround** Sound, everyone can hear the stereo effect because there is a larger sweet spot.

42   Howard W. Sams & Company **Complete Guide to Audio**

SURROUND SOUND COMPOSITION

Because surround sound utilizes 3 to 6 channels, you will need more speakers than usual, as well as a way to amplify the signals sent to them. The soundtrack is coded with the surround sound information. This means you will need a special decoder to listen to the multiple channels. The Dolby surround sound decoder is usually built into an A/V receiver. Lastly, you will need to know how to set up the speakers, as each is placed in a specific location within your viewing room.

DOLBY SURROUND SOUND

Dolby surround sound was released for the home market in 1982. It encodes 3 channels onto a film or movie track: left, right and surround. In order to hear this signal, hook your surround sound receiver up to two front tower speakers for the left and right channels. Two tiny speakers are then placed at the sides of the room for the remaining single channel.

Dolby Laboratories sparked the roaring flames of the home theater market with the introduction of Dolby surround sound. But what exactly is surround sound? The Dolby process codes several channels onto a movie soundtrack during filming or editing. You may remember from Chapter 3 that each channel is a separate recording. Each recording is then played back through one or more speakers that are placed apart from each other. This creates a superb sound stage that encompasses your entire body; something a regular two-channel stereo cannot do.

The difference between surround sound and a regular system is that with surround, there are now speakers placed at the sides in addition to the regular twin speakers in front of you. The front speakers are two channels. The back speakers combine to form a third channel. We will describe what function each channel and speaker plays in a moment.

DOLBY PRO LOGIC SURROUND SOUND

Dolby Pro Logic was the successor to surround sound. It was released on 1987 receivers, and is still being used for low- to mid-end units and movies. See Figure 4-2. Pro Logic has 4 channels encoded onto the movie track: left, right, surround, and now center, or the dialog channel. This channel is used to make it seem like the voices are right in front of you. In order to hear a Pro Logic signal, hook up your Pro Logic-equipped receiver to two front speakers, the two tiny speakers for the surround channel, and a specially-shielded center speaker to set atop your TV. This will make the sweet spot obsolete, as almost anyone in the room can now listen to the full sound stage.

DOLBY DIGITAL: AC-3

Dolby Digital is virtual sound. When new digital television standards and new digital player devices were introduced, so was AC-3 for the home. This is 6-channel surround sound, or more accurately, 5.1 channels. This means there are 5 channels and a dedicated subwoofer signal. (Don't ask me why Dolby doesn't just say "6 channels!") See Figure 4-3. This includes left, right, center, subwoofer, and now (instead of the surround sound being mono) a left surround and right surround channel. The only speaker configuration difference is that the rears are wired independently to their separate channels, and there is a line for a subwoofer. Dolby Digital signals can only be received by a DVD, DTV, and soon from a direct broadcast satellite (DBS), because the signal is overpacked with sound information. So, if you want to utilize that new Dolby Digital receiver, you will need a DVD player/laser disk player that supports it, or wait until the digital TV stations or DBS are on line.

NOTE: Dolby Digital is the consumer name for Dolby AC-3. They are often used synonymously; but AC-3 is actually the data reduction scheme.

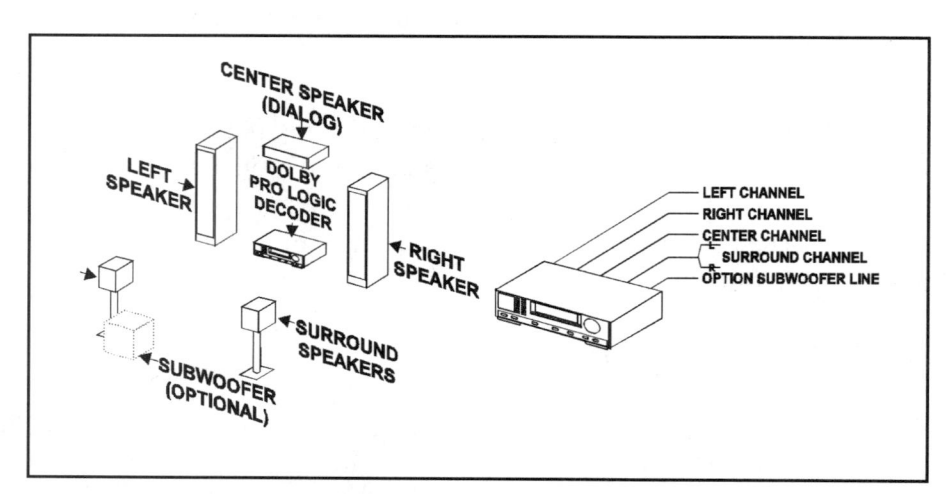

Figure 4-2. Dolby Pro Logic surround sound.

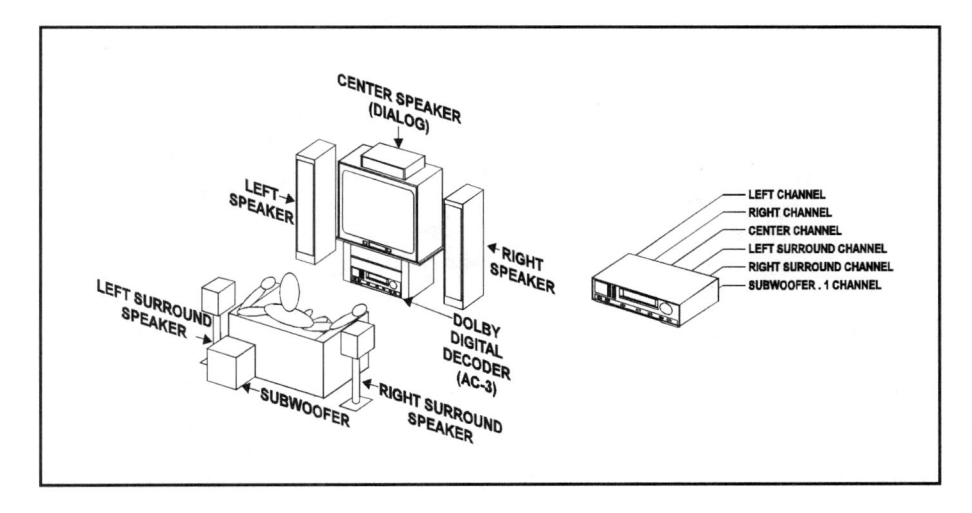

Figure 4-3. Dolby Digital surround sound.

HOW SURROUND SOUND IS ENCODED

While a surround sound TV show or movie is in production, four audio channels are concurrently recorded. Dolby circuitry is then used to encode them. This makes the 4 channels able to be carried on the regular left and right channel alone. The process is called *surround matrix*. See Figure 4-4.

A Dolby Pro Logic decoder is then used to recover the encoded signal information (which was sent to it as a combined lft./rt. stereo pair). This makes the original movie track's four channels available to the rest of the stereo's equipment.

EQUIPMENT

One of the most-asked questions when dealing with basic audio is, "What do I need for surround sound?" You will need a signal source, a receiver with a built-in decoder, and speakers. Check out the next few sections for more explanation.

SURROUND SOUND SIGNAL SOURCE

In order to listen to a Dolby surround sound movie, you (of course) need Dolby-equipped audio products. However, you also need to re-

Chapter 4: Home Theater Sound

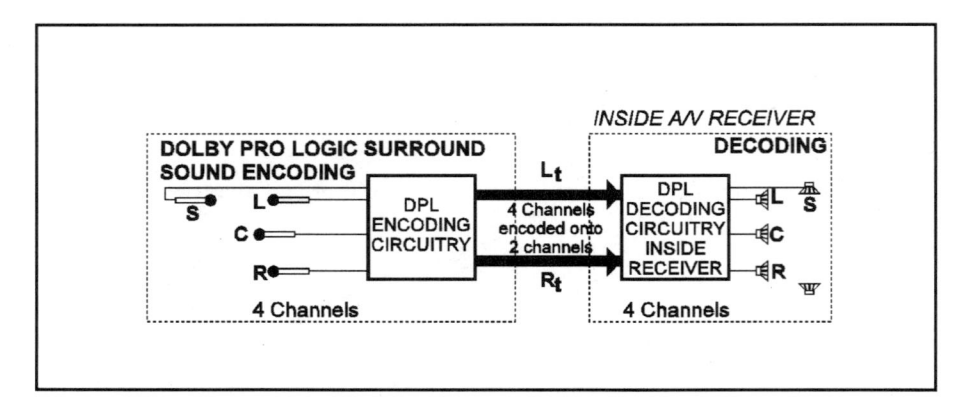

Figure 4-4. Dolby surround sound encoding/decoding.

ceive a Dolby surround signal so the decoder can do its job. Some TV shows, such as *The X-Files*, are transmitted in surround sound stereo. Most VCRs can play surround sound VHS tapes. Here is a full rundown of other signal sources you can use to receive a ring of sound:

DOLBY PRO LOGIC SOURCES

Stereo TV Signals: As we said, some TV shows are transmitted with the Dolby Pro Logic signal encoded into the MTS stereo signal. A Pro Logic-equipped A/V receiver will be able to decode and send your favorite TV show's audio to your surround speakers. Beware! Some broadcasters trash the stereo signal, leaving you with a bland mono soundtrack.

VCR: If you look, somewhere on the videotape's box, there should be a Dolby DPL-DD logo. This means the VHS tape has a Dolby Pro Logic-encoded soundtrack. If not, well, at least the regular stereo mode should work. Maybe.

DBS: Because digital satellites can compress video and audio signals, they are able to send more information in a smaller amount of space. This means that you can receive surround sound movie tracks as well as 30+ channels of digital-quality music.

Cable: Some cable companies can send a Pro Logic signal. You will have to check with your local cable provider.

DOLBY DIGITAL (AC-3) SOURCES

DVD: Digital video disc players are the hottest-selling item in the consumer A/V market. They can spin out a Dolby Pro Logic or Dolby Digital soundtrack. Aside from a small laser disc player market, this is currently the only signal source that provides Dolby Digital surround sound.

Future — DTV: The new digital television broadcast standards include provisions for DD.

Laser Video Disc: Some laser disc players are able to make use of Dolby Digital flicks.

Future — DBS and Cable: Once manufacturers jam the needed Dolby Digital circuitry into DBS equipment, then AC-3 will be available. Maybe by the end of 1998, AC-3 will show its face. Cable companies are likely to follow.

RECEIVERS AND DECODERS

Now that you understand the need for a surround sound signal source, let's look at the rest of the contraptions you have to deal with. The A/V receiver is the base from which most home theaters are built. It contains the Dolby decoding circuitry, preamp, power amp and tuner. Some units have the decoder as a separate unit or built into a television, DVD or other piece of equipment. Either way, you still need an A/V receiver to control and amplify 5+ speakers.

Pro Logic: These are the most popular A/V receivers for the no-hassle hookup consumer market. They are usually an all-in-one package with sufficient wattage to power your speakers. Typical options include some type of "ambiance" switch, which makes the sound seem like it is in altered environments, such as a concert hall or a jazz club. Your tuner will likely be of the digital variety; able to lock out any station with the push of a button. The receiver should be able to handle multiple input sources, such as a tuner, VCR, cable, DBS, DVD, laser disc, CD player, etc. There should be outputs for standard video, composite video, S-video, digital (if possible), speakers, and a subwoofer signal line. Most receivers come with a remote control.

Dolby Digital: Digital receivers have not quite hit the market with strength yet because there are only a few ways to receive a DD signal. However, if you plan to expand later to DTV and DVD, the option is worth it now. There are receivers on the market right now that are touted as "Dolby Digital Ready." This simply means that there are 6 channels available. However, this does not necessarily mean that there is a DD decoder built in. Typically, there should be a Pro Logic decoder present. If you have a DVD player, the decoder is likely built into it. Eventually, the DD decoder will be built into all receivers along with a Pro Logic decoder; so we will be able to plug in any DD device and send out 6 channels of virtual sound to the speakers.

SPEAKERS

NOTE: See Chapter 8 for more information on speakers.

Building a home theater brick by brick is a breeze. Once you have a nice receiver with the regular complements, it's time to consider the speaker selections. If you want, you can start with only the two front pairs and later add the center, surrounds and subwoofer. However, most audio stores now offer a set of five speakers as a package. If you are not too particular on choices and want to leave them in the hands of the manufacturer, go for a package. If you want choice and expansion ability, buy each set according to what you think sounds best.

HINT: Don't always buy for the brand name, but rather how the speakers sound. I have heard a $600 set of brand name speakers that sounded like someone was screeching in my ears; yet a cheaper set of consumer speakers sounded heavenly.

Front Pair: A normal set of tower speakers is fine unless you are going for the subwoofer/satellite combo. See Photo 4-1. Try to get something that will handle the recommended wattage of your amp. If you are nondiscerning and simply want *sound*, go for at least 50 watts per speaker. Average is 100-200 watts. (I personally would never go under 150 watts per channel for the left, right or center for an average-size room).

Center: You will need a shielded speaker that can handle the wattage of your amp as well. Try to match the wattage with the left and front speakers: 100W left, 100W right and 100W center.

Surround Pair: Whatever wattage is used for one channel in the left, right or center is usually split between the two small surround speakers. I personally recommend the highest power in the surround channel as you can afford, because there is an audible difference in cheaper, low-power speakers/amps. Actually, it's a sickening difference.

Subwoofer: Passive (powered by your receiver's amp) subwoofers are rare unless the amplifier is a separate component from the receiver. This also is not recommended because of having to run such a huge cable around the room to power the sub. Instead, look for an internally-powered *active* subwoofer with as high a wattage as you can afford. If you are not a bass head, put the extra $400 into a better trio of speakers up front.

Photo 4-1. 130 watt floor standing speakers. Reproduced with the permission of Sony of Canada.

See Chapter 8 for a primer on how to set up your speakers correctly.

AMPLIFIERS

See the next section for amplifier explanations and recommendations. Remember that the higher the wattage, the higher the volume. This also means better-quality sound at lower listening levels. For this reason, always go with as much total wattage or watts-per-channel as your budget allows.

CAUTION: Do not set a speaker that isn't specially designed with shielding on top of your TV set. The magnet inside can damage your television's picture tube.

DTS AND OTHER NON-DOLBY SURROUND SOUND SYSTEMS

There are other multichannel surround sound systems on the market in addition to Dolby's choices: digital theater systems (DTS), Sony Dynamic digital sound (SDDS), and others. Let's not get into the explanation of how each differs, except to say that they are not compatible with each other. Most movie tracks use some type of Dolby surround sound

THX

This has to be one of the most misunderstood, misused terms in A/V. THX is similar to Dolby in that they both do not actually produce any salable products. The THX logo on A/V equipment means that the electronic component or speaker has been certified by THX to have met THX standards.

The main purpose of a THX-certified component is to allow you to reproduce the sound quality and character of the mixing stage where a movie's soundtrack was originally recorded and mixed. In other words, THX equipment is needed to reproduce what the director or sound engineer wanted you to hear, along with all its spatial information, quirks, acoustics, *everything*! It's *exactly* what they intended you to hear. So, audio equipment that is THX certified will recreate exactly what the director heard while mixing the movie. This is not possible with run-of-the-mill receivers and speakers.

Home THX is different from a movie theater's THX. Because the acoustics of a small living room are obviously different from those of, say, Mann's Chinese Theater, THX engineers have figured out ways to compensate electronically and by setting various standards. Home THX is the result: the same quality sound of a THX-certified theater, inside your home. For more information on THX, and a list of THX-certified home theater components, see their Website at WWW.THX.COM.

NOTE: THX is not a replacement for surround sound technology, such as Dolby Pro Logic or Dolby Digital. It is merely an enhancement and standard.

(usually Pro Logic). So, it seems to be in your best interest to stick with Dolby, unless you can find equipment (such as DVD players and receivers) with more than one surround sound standard built in.

WRAP UP

Remember that sound is half the experience of watching your favorite movie at home. Do your family and yourself a favor and build some type of surround sound system from the start. Don't give into the temptation to spend less on a regular 2-channel hi-fi, or you will always wonder, "Would this flick sound better in surround sound?" IT ALWAYS DOES!

CHAPTER 5
AMPLIFIERS & PREAMPLIFIERS

"Rock and Roll!" Michael J. Fox as Marty, *Back to the Future*

Studying sound is more palatable when you apply it to a more practical subject. I am talking about the acoustical application of sound — audio equipment! Let's take a deeper look into each component, starting with *amplifiers*. This is often one of the weakest link in the audio chain.

NOTE: AMPLIFIER and AMP are synonymous.

The word *amplifier* comes from *amplify*, which means, "To make stronger." In the electronic sense, it means to strengthen an electrical signal with an amplifier. Audio equipment uses electronic gizmos called *transistors* (or vacuum tubes) to accomplish amplification. What exactly is being amplified? The signal from a microphone is amplified. The faint signal coming from a record needle swinging back and forth in an LP's groove is amplified. The loudspeakers are sent an amplified electrical signal that contains the sound your ears are waiting patiently to hear. See Figure 5-1.

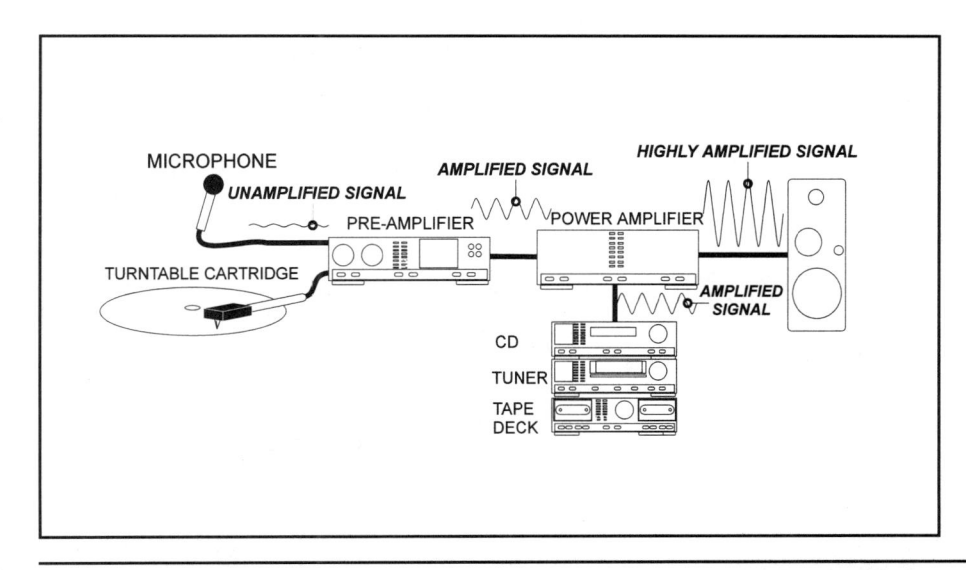

Figure 5-1. Amplification in audio equipment.

Chapter 5: Amplifiers & Preamplifiers

HOW AN AMPLIFIER WORKS

An amplifier *amplifies*. It increases the electricity to levels a speaker can use. The part of the electronic signal that is actually amplified in an audio system is typically the amplitude of the wave, also known as *signal level*. See Figure 5-2. It's like an automobile's power steering: it takes your exact input, adds torque (energy), and outputs it to the wheels.

In an audio system, the hands on the wheel may be a person speaking into a microphone. The microphone changes the sound energy into electrical energy. However, the energy is nowhere near enough to vibrate a speaker cone back and forth. Therefore, the signal level from the microphone is wired through an amplifier. This increases the amplitude of the voice signal (now electronic) to levels the speakers can use. If you want to make your voice louder without raising it, you simply raise the amplitude going to the speakers further (amplify the signal level). No straining your voice — let the electrons in the amplifier do all the work.

AUDIO AMPLIFIERS

Modern audio equipment uses two types of amplifiers. The preamplifier is used to receive signals and boost microphone or LP needle signals, and to control such things as volume, level, etc. The power amplifier or output amplifier is used to pep up weak signals to levels that are capable of driving a loudspeaker. Think of the preamp as the brains and the output amp as the brawn.

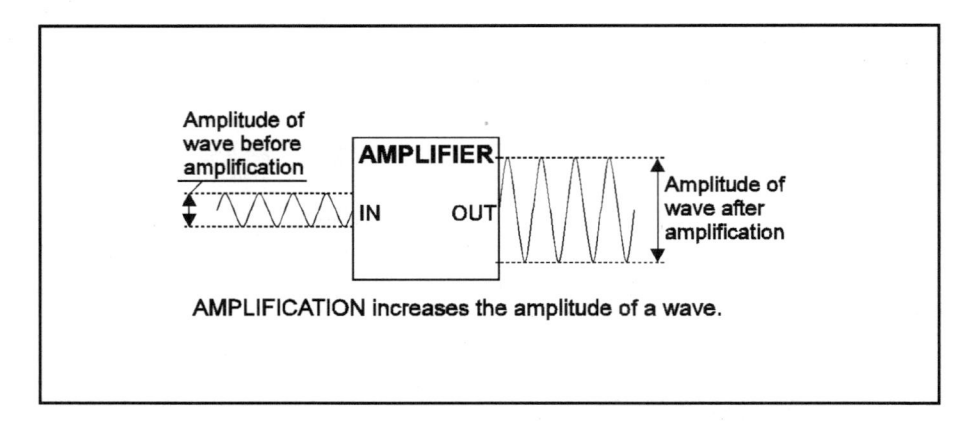

Figure 5-2. Amplification.

AMPLIFICATION increases the amplitude of a wave.

PREAMPLIFIERS

The word *preamplifier* is a misnomer because its purpose is more than preliminary amplification. Most modern audio components, such as CD players and tape decks, don't need the signal preamplified. *Control amplifier* would be a more appropriate term since it is controlling the various signal levels, volume, etc.

WORKINGS

The purpose of the preamp is to increase the initial signal level of such things as a phonograph cartridge or a microphone's output. This allows the signal to be processed further without appreciable degradation in quality.

The control section of the preamp handles:

Level Control: Volume! You have to be able to control the volume on your system. This is also the preamp's job. Some CD players have built-in volume controls that replace the preamp's functions.

Balance Control: This controls the balance between the speakers. If the speakers sound heavy on the left side, turn the balance to the right until they are equal. This is important in order to achieve optimum stereo imaging.

Bass and Treble Control: These two controls let you adjust the level of the bass frequencies (lows) and treble frequencies (highs). Some people prefer a hotter high frequency, and others want those pounding low-bass frequencies to be stronger. It's a matter of taste.

Equalization: Yes, we are talking about those mysterious sliding levers on some receiver faces. Do you *really* know what they do besides provide a high coolness factor? They basically replace the treble and bass knobs, giving you more control of the system's various frequency bands.

Signal Source Control: The amplifier needs to know which signal source to amplify. This is no longer an A-to-B task with many audio compo-

Chapter 5: Amplifiers & Preamplifiers

nents: CD players, tape decks, AM/FM/TV tuners, home theater components, etc. So, the preamp also acts as a signal source control. It is the traffic director of the system.

Mixing: The preamplifier provides a train-station-type of routing system to control components used in mixing music. For example, setting the input signal to *CD* and the output to *tape deck* will allow you to record a CD.

Other Controls: On some systems, a "power-up sequence" is required. One component turns and then another, then another... This prevents damage to certain sensitive items such as speaker drivers.

Phantom Power to a Microphone: Some microphones require power be routed to them from the preamp. A tape deck has this feature built in, if it is not already included in your receiver.

HOW PREAMPS ARE USED IN A MODERN STEREO

A separate preamp is not necessary. Most receivers have a built-in control amp and phono cartridge input. As well, most tape decks have a microphone preamp built in. CD players, tape decks and receiver signals do not need a boost from the preamp, but only need to be controlled.

Do you need a preamp? If you are going to run a phonograph, it is a requirement. Some CD players can all but eliminate a preamp or control amp, as they have variable level outputs that can be connected directly to a power amplifier. However, you will loose the ability to use one volume control for all components. Thus, it is best to have a preamp built into the circuitry; usually the receiver. After all, the preamp *is* the brains of the operation.

POWER (OUTPUT) AMPLIFIERS

Now for the muscles of the audio equipment. Your stereo flexes its biceps by pumping out power to the loudspeakers. The more elevated the wattage, the greater the volume, and the greater the impact upon the listeners.

Not only does the power amp require potent power levels, but the signal being amplified has to be as free of externally irritating noise as possible. This noise is more accurately referred to as the "signal-to-noise ratio." The more signal retained from the start of the electronics chain until it reaches the speakers, the cleaner the system will sound. The amplifier is typically the "mud puddle" in the electronic signal's path.

WORKINGS

A power amplifier injects mass amounts of electrical power to a sound signal. This is done with electronic components. One type of system uses vacuum tubes, and the other, more popular system uses solid-state transistors. The transistors or vacuum tubes form a network that takes in an electronic signal, adds power to it by increasing its amplitude, then routes it to your speakers. A preamp works in the same fashion, just on a smaller level.

THE MISCONCEPTION OF WATTS

One of my favorite movie scenes is from *Back to the Future*. Marty is at Doc's laboratory cranking the amplifier equipment to the max and plugs in his guitar. With one strum of the string, WHAAAM! He is thrown against the back wall. This scene, plus the heavy metal mentality of the 1980s, created the biggest misconception about amps today. More power does not equal better sound.

An amplifier's output is rated in watts (W). This is a measurement of the power the amp can create. Most people wrongly choose an amplifier based on this number, thinking the higher the watts, the better-sounding the system. Not true. You should look for an amplifier that produces clean sound with no distortion or weird characteristics. A 40W NAD amp may sound better than a 400W Pioneer.

VARIOUS TYPES OF AMPLIFIERS

Originally, amplifiers were large cones that acted as a mechanical amp/speaker. Eventually, glowing vacuum tubes provided an electronic means of amplifying a sound signal. The problem was that they ate loads of

NOTE: Try to go for the highest wattage rating possible, even if you are not going to listen at house-crumbling volumes. The higher the wattage, the cleaner the sound at lower power levels. But never forsake a good-sounding amp because of its lower power rating.

Chapter 5: Amplifiers & Preamplifiers

power. The discovery of the solid-state transistor changed audio technology forever, as it was possible to provide a clean amplified signal without wasting juice. For some reason, vacuum tubes have yet to flicker out in the market as they provide more than just amplification in some audiophile eyes (or ears). I am talking about nostalgic-sounding equipment. Here is a description of each amplification method.

VACUUM TUBE AMP

Audio tubes are a high-end market. Typically, a quality vacuum amp will punch a several thousand-dollar hole in your wallet. But some people think the investment in venerable audio equipment is priceless.

Pros: Very classic-looking, especially when the amp is designed to display the tubes themselves. Provides a less "transistorish" sound (there's more character to it).

Cons: Power robbers. Tubes burn out, and are more susceptible to hum and vibration-induced interference. It's hard to replace certain classic tubes.

SOLID-STATE TRANSISTOR AMP

Pros: Reliable; economical on power; cheap.

Cons: Some people don't like the plastic-sounding characteristics. High power loads damage the transistors quickly.

WHAT KIND OF AMP DO YOU NEED?

Tube or Transistor: If you prefer that tubelike quality and the shear beauty of the glowing glass, go for a tube amp. Expect to pay a small fortune. More likely, you are going to purchase a receiver with a built-in power amp and preamp. They are invariably transistor models. If you have grown up with transistorized audio equipment, you will never miss the tubes. Stick with the cheaper mass-market receivers.

Speaker Matching: Always keep in mind the type of speakers you will be using as well. The amplifier is going to work hand-in-hand with them to bring a distortionless audio presentation to your hungry ears. Make sure the speakers and amp work well together. This includes matching impedance, obeying the polarity, and not overdriving them.

Watts: The types of speakers to which your amp will be providing power should determine the amp's power rating.

NOISE & DISTORTION

All sorts of electrical noise can creep into an amplifier. Creepy alien sounds that just don't belong are then passed onto your speakers. This can cause anything from an irritating passage in your favorite CD to the destruction of your speakers, if the amp sends out a corrosive square wave. Always listen carefully for quirky noises and distortion in an amplifier by putting it through its paces before a purchase.

CHANNEL SEPARATION

As you learned in Chapter 3, it is important for the sounds from the left and right channels to be totally independent of each other. They are completely different recordings that, if mixed, will destroy the stereo effect. So the amplifier must separate the channels as much as possible.

IMPEDANCE MATCHING

A speaker's impedance (resistance or opposition to an alternating current) is measured in ohms. Most speakers and power amps have either a 4-, 6- or 8-ohm impedance. 6 and 8 ohms are the most common, and 4 ohms are typically on high-cost equipment. The lower the impedance, the higher the current being drawn from the power amplifier. Most receivers have recommended impedance for your speakers. However, almost any amplifier can drive any speaker load as long as you don't raise the volume too high. You can use a 4-ohm speaker in place of an 8-ohm one under several conditions. Take caution when doing this:

Do not crank the volume to ear-shattering (or speaker-blowing) levels. You can add a high-power 4-ohm resistor in series with a 4-ohm speaker to bring the impedance to 8 ohms. Beware! This may add noise to the speaker.
Wire two 4-ohm speakers in series on the same channel.

How can you tell if things are fine to run at these levels? Put your hand on the top of the receiver or amplifier. If it is way too hot, turn down the volume or match the recommended impedance.

NOTE: If you already own a set of speakers and are replacing an amp, make sure the minimum wattage required for your speakers compares to those of the amp (usually around 10 to 50 watts). Maximum wattage of the amp is not as important. You can always control the volume.

HOW MANUFACTURERS MEASURE THE WATTAGE OF A POWER AMP

Most manufacturers rate the wattage of their receivers in one of two ways. Some are rated as "watts-per-channel," and others are rated as total wattage available to all channels. See Figure 5-3.

EXAMPLES

Watts-Per-Channel: 100 watts for the left channel, 100 watts for the right channel, 100 watts for the center channel, and 100 watts for the two surround sound speakers.

Total Wattage: 100 watts x 4 channels, or 400 watts total. This means that each channel can draw as much juice as it needs.

A receiver with *N* watts x 4 channels is usually a cheaper mass-market unit. Some people think that the total wattage sometimes distorts the music. Let's say that there is a section where a cymbal is crashing, a bass drum in thundering, and a violin is harmonizing, all at once. This can easily overdraw the total wattage and cause distortion. Higher-cost Dolby Digital and Pro Logic units have 100 to 500 watts per channel, up to 5 channels. This makes for better separation and less power

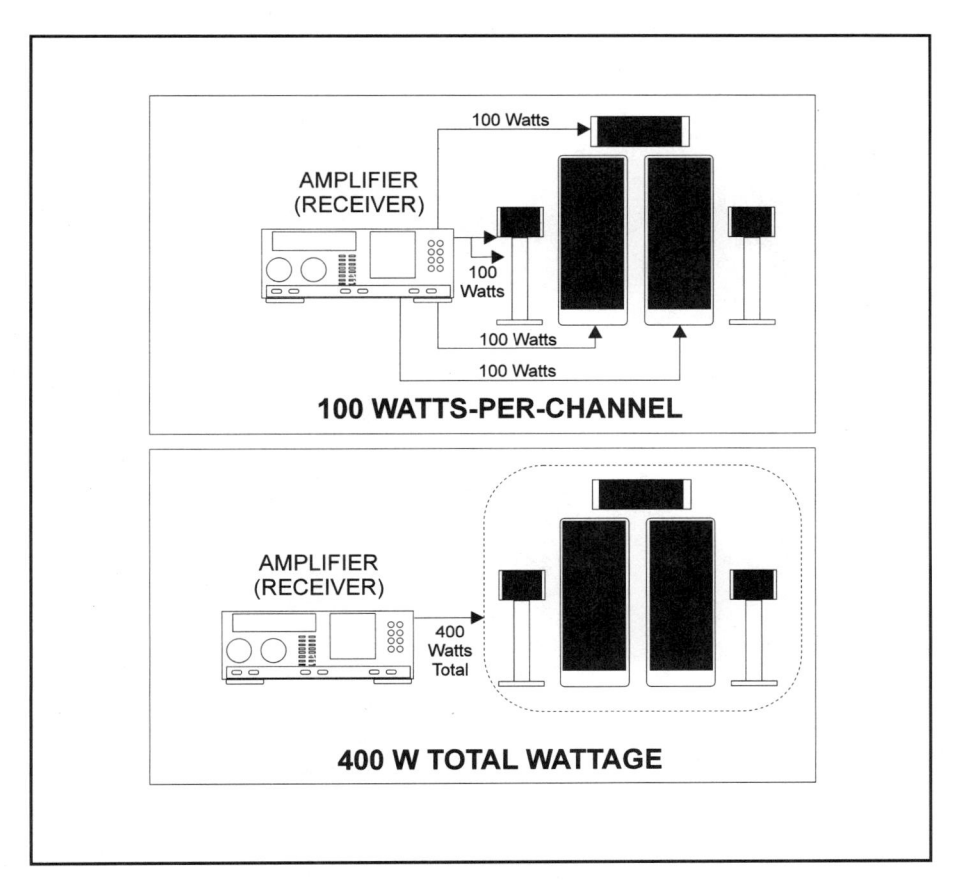

Figure 5-3. How manufacturers measure the wattage of a power amp.

problems in intense sections of the music. Which one you choose depends on your preferences and pocketbook.

WRAP UP

In the run-of-the-mill audio market, preamplifiers and power amplifiers are bundled into the receiver. In the next section, we will take a further look at amplifiers as they relate to the receiver, and give you more amplifier purchasing advice in Chapter 10.

Chapter 5: Amplifiers & Preamplifiers

CHAPTER 6
RECEIVERS & SURROUND SOUND DECODERS

"Music finds its way where the rays of the sun cannot penetrate."
Soren Kierkegaard

Receiving, distributing, controlling and amplifying a stereo's signal is the thankless composite task of the stereo receiver. This overdriven piece of audio equipment usually consists of a tuner, preamplifier and power amplifier, and a place to plug in your various audio and video sources and outputs. See Figure 6-1. The receiver is the brains and taskmaster of the entire operation. In addition, an audio/visual (A/V) receiver contains the decoder hardware that will take a Dolby surround signal and turn it into 3 to 6 channels of surround sound pleasure. Most new receivers also have a microprocessor-type device called a *digital signal processor* (DSP). This will warp the sound into various settings such as a jazz hall or concert arena.

In this chapter, we will study the function and application of tuners and decoders. The preamplifier and power amplifier were covered in the previous chapter. Now we'll look at the receiver as an integral component and make recommendations. Pay attention! This section is important. The receiver is the mastermind, and you don't want a cheap and useless receiver looking after the rest of your valuable audio equipment.

THE TUNER

A receiver contains the circuitry that tunes into the electromagnetic radio waves that surround us. Once locked onto a certain frequency or

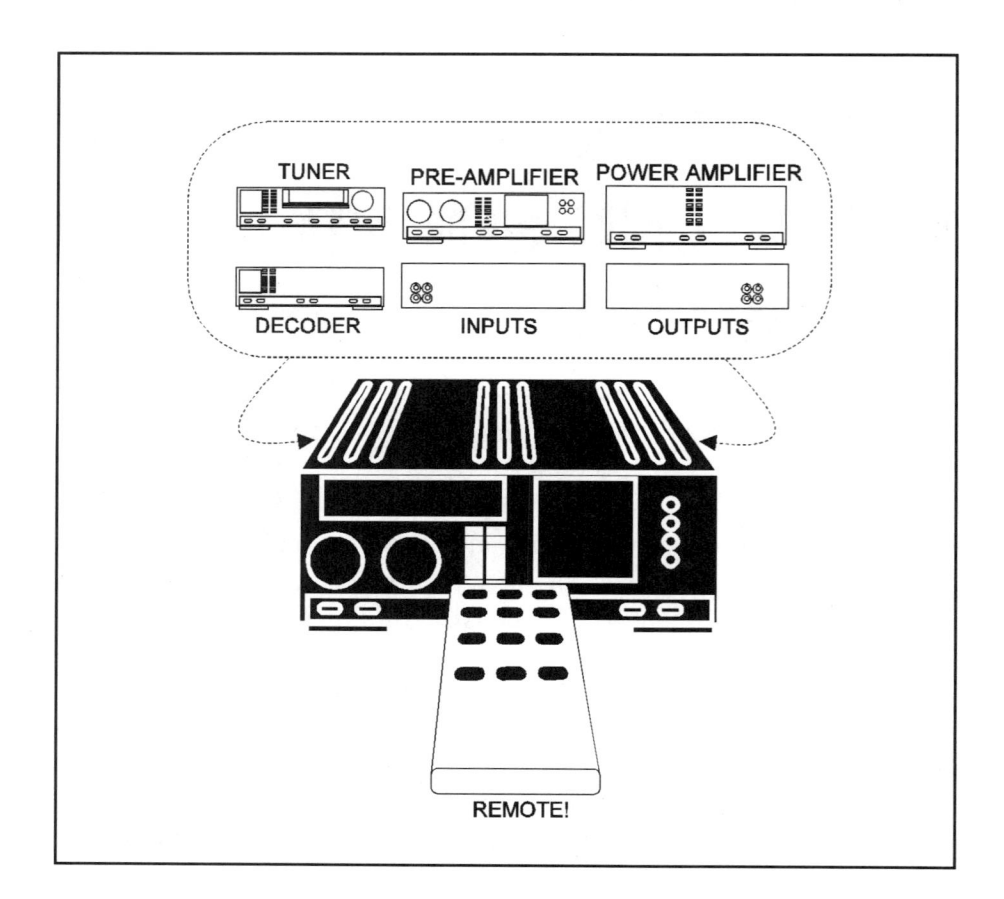

Figure 6-1. Innards of a modern receiver.

frequency band, the circuitry will receive a signal from a radio station or TV station and turn it into audio. This circuitry is called the *tuner*.

RADIO WAVES

Your tuner receives AM/FM signals which contain sound information. A higher frequency is used to carry the lower sound frequencies. This is called *modulation*. When a tuner receives the signal, it is then demodulated. The higher frequency carrier wave is discarded. See Figure 6-2. There are two types of modulation/demodulation functions used in most tuners, currently.

Amplitude Modulation: Audio frequencies are modulated over a radio frequency of about 540,000 Hz (540 kHz), terminating around

62 Howard W. Sams & Company **Complete Guide to Audio**

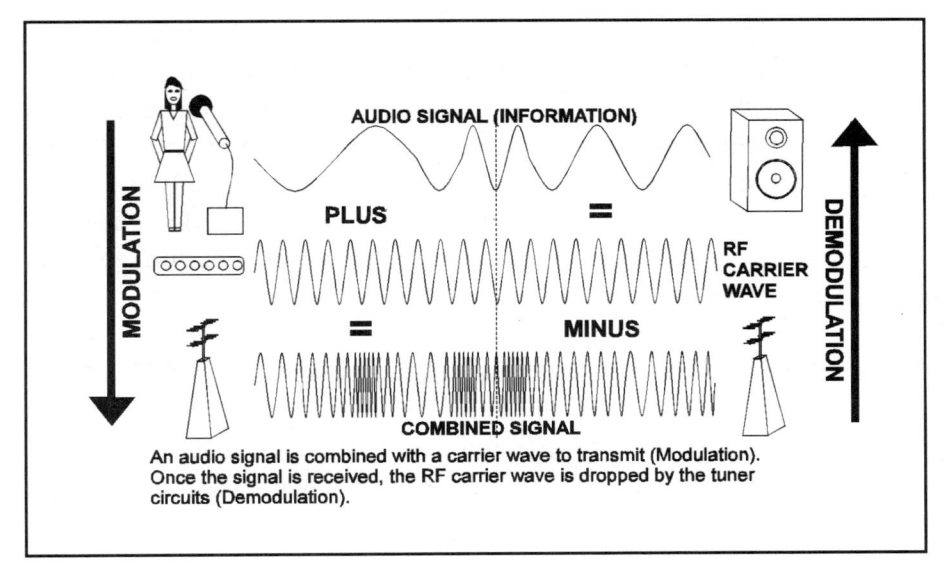

Figure 6-2. Modulation and demodulation of an audio signal.

1,700,000 Hz (1700 kHz). The sound information is laid into the modulated frequency in the form of varying amplitudes in the wave.

AM is prone to destructive noise from sources such as electric motors, car ignitions, and other devices that produce a spark. For this reason, AM is very limited and low-fidelity, and therefore rarely used these days for audio applications. However, AM stations can transmit farther. So, if you live in a fringe area, you can still think you are part of the civilized world by receiving AM radio stations.

Frequency Modulation: FM takes the original sound frequencies and modulates them onto a frequency band about 200 kHz wide. The audio FM radio band starts around 88 megahertz and extends to 108 MHz. This gives us about 100 channels of FM. Television is also transmitted using FM waves, but carries comparatively vast amounts of picture and sound information. FM can transmit a dual stereo signal but is unable to extend vary far. A roof antenna may be needed to receive the strongest FM signals for your receiver.

HOW A TUNER WORKS

A tuner uses an electronic circuit called a *heterodyne receiver*. The device receives a certain radio frequency or frequency band, discards

Chapter 6: Receivers & Surround Sound Decoders

the carrier frequency and leaves the original sound signal, which is then sent to the preamplifier. When you turn the AM/FM dial on a tuner, you are actually setting the frequency you want the heterodyne receiver to receive, and determining which modulated frequency is to be thrown away. Most tuners, including TV tuners, now electronically lock onto the strongest frequencies automatically. This way you don't have to futz with a dial and fine tune for hours; simply push a button, and you can listen to your favorite radio station.

RECEIVING THE STRONGEST, LEAST NOISY FM SIGNAL

The tuner needs to be fed a healthy supply of radio waves, sent to it through an antenna. The antenna receives a weakened signal from a radio station and sends it down a wire to the tuner. The tuner then locks onto the correct frequencies and amplifies the signal so the rest of the audio equipment can make use of it.

Even though an antenna receives a relatively feeble FM wave, it is best to make sure it is getting the strongest possible signal. Remember that a signal amplifier can only amplify the signal it is receiving. If the signal from the antenna is noisy to begin with, or weakens, the amplifier will only make it worse: it amplifies the noise. You must first strengthen the original incoming signal by using a larger, more directional antenna: not a stronger signal amplifier.

ANTENNAS

The best antenna to use is a house-mounted TV/FM antenna. Most antennas that are perched on roofs are of this type, and can be used to receive an incredibly strong FM signal. You will need a signal splitter, as pictured in Figure 6-3. Connect a 300-ohm flat cable to the FM screws on the splitter and run them to the FM antenna inputs on your receiver. Some components have this built into the circuitry, eliminating the need for a splitter.

HINT: If you are having trouble receiving a clean signal from a radio station, raise your antenna higher or get a stronger model.

The next choice would be to use the antenna that came with your receiver. It is usually a wire that is strung across a wall or cabinet. If you are relatively close to a radio station, this antenna should work fine.

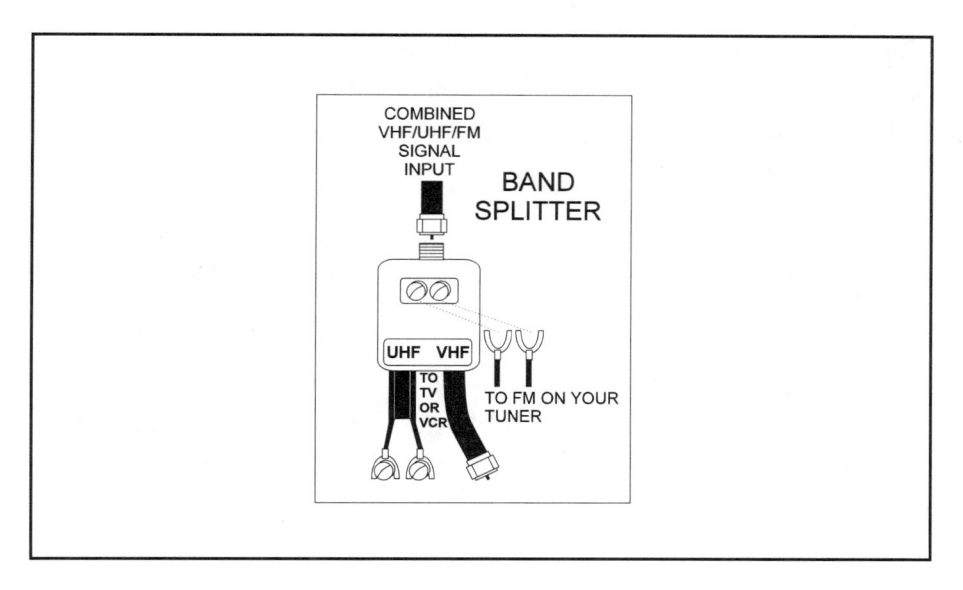

Figure 6-3.
Band splitter.

DIGITAL AUDIO BROADCAST (DAB)

FM radio will soon be replaced by a new standard, currently dubbed digital audio broadcast (DAB). This will be a purely digital signal that radio stations around the world will fling out to our stereo receivers. Digital television is nearly here as well. What this means is that you will receive a CD-quality audio broadcast or extremely high-quality A/V signal. The other advantage is that the radio stations can offer loads more channels using audio compression technology. There is no degradation of the signal, either. You either receive perfect sound or none at all. I guess this follows the nature of digital: ON-OFF-ON...

What does this mean to you as an audio or video consumer? Yes, once more you must upgrade to a digital audio/video receiver once DAB standards are implemented. Not to worry, though; there will likely be a period of a few years in which analog FM and digital FM are simulcast.

THE SURROUND SOUND DECODER

Audio/visual (A/V) receivers are the fastest-growing segment of the receiver market. They contain the circuitry that decodes the Dolby surround sound signal into the coolest sound stages or movie stages you've

Chapter 6: Receivers & Surround Sound Decoders

ever heard. An A/V receiver is highly recommended as your first receiver purchase. You can expand your home theater from this base without having to purchase new equipment. Most receivers on the market today are of this variety.

A/V receivers currently come in three similar species. One is standard surround sound. Next is Dolby Pro Logic. And last is Dolby Digital. See Chapter 1, section *25 Audio Terms You Should Know*, and Chapter 4 for a further explanation of each.

Pro Logic is the popular choice right now. Most low- to mid-end receivers have this circuitry. However, Dolby Digital *is* the future standard and will be on more midrange systems soon. For now, DD is a high-end techno treat.

DSPs

DSP means *digital signal processor*. In a nutshell, it is a microprocessor that is able to process audio or video signals efficiently and according to how the engineers command it. It is the *ambience* button on your A/V receiver at work. It allows you to warp the sound stage into any configuration you want. Maybe you want a favorite piece of music to mimic the reverberations of a famous concert hall. But now you want it to sound like the musicians are right in that small room with you giving a personal concert. With the flip of a switch on the receiver, the DSP changes the characteristics of the sounds being sent to your speakers, recreating those sound scenarios.

REMOTE CONTROLS

Nearly all rack system receivers, small combination systems and separate receivers come standard with a remote control. See Photo 6-1. It works by sending out pulses of infrared light signals that the receiver then decodes to perform the demanded instruction. If you are not familiar with highly-technical controls, try to go for a unit that has a simple remote. Make sure the remote that comes with the receiver can be programmed into a universal remote control for A/V purposes. Otherwise, you will be stuck using five different remotes for your entertainment system!

REAL-WORLD RECEIVERS

Most receivers on the market today contain:

- Some type of surround sound option, usually Pro Logic. See Photo 6-2.
- A preamp that will handle a phonograph if needed.
- A fair-sized power amp. Go for at least 60 watts each for left, center and right, and about 20 watts for each surround.
- AM/FM tuner (possibly a TV tuner) with 20 to 40 presets and scan feature.
- Remote control with as many gadgets as your mind can absorb without having a brain freeze.
- Bass/treble controls or equalizer for adjusting.
- Inputs and outputs for all your additional equipment. Most have audio, video and digital I/Os.

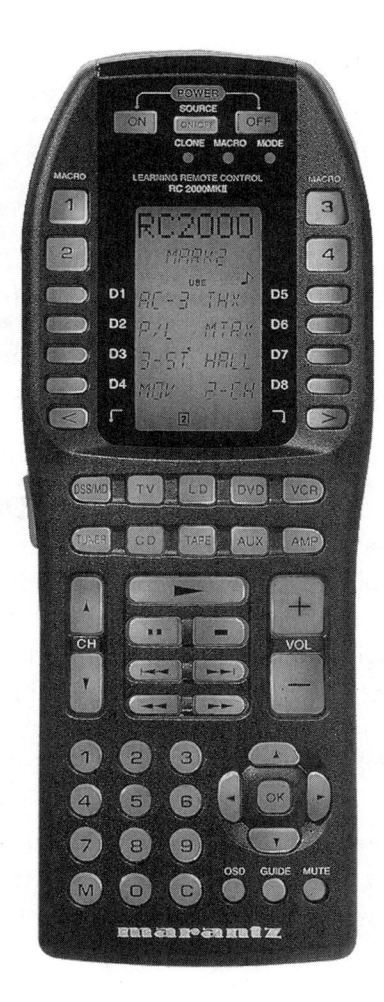

Photo 6-1. Receiver universal programmable remote control. Reproduced with the permission of Lenbrook Industries, Marantz Canada.

WHAT TO LOOK FOR IN A RECEIVER

In addition to the recommendations for amplifiers and preamps in the previous chapter, here are a few more:

Pro Logic is likely to be the de facto until the end of the century at least.

Always choose an A/V receiver. You may think you will never need it, but you will in the future. They are only a few bucks more for a Pro

Photo 6-2. Hi-fi receiver with Dolby Pro Logic decoder. Reproduced with the permission of Sony of Canada.

Logic model, and the Dolby Digital models soon will follow step. You have to purchase extra speakers, but you can always buy only the front speakers first and use the Dolby 3 Stereo mode for now. When you are ready, you can add a set of side satellite speakers and a hard-hitting subwoofer. The in-your-face sound experience is a thrill, so go for the A/V receiver now and avoid having to buy one later.

WHAT TO AVOID

Stay away from older surround sound receivers. Purchase a Pro Logic or Dolby Digital receiver. If you are purchasing used equipment, be aware that the amp is usually the first component to go. This is due to the high currents being force-fed to its transistors and capacitors. With age, caps dry out and cause all sorts of noisy problems and other audio headaches. Because the amp is an integral part of most receivers, replacing an amp means replacing the whole receiver. It is possible to buy the amps separate from the decoder and tuner, but not recommended for the novice.

Try to avoid complex controls if you consider yourself a technologically challenged individual. Those thousand points of light on the receiver panel may make you look like a nuclear engineer at the controls; but if you don't know how to adjust them, there can be a sound quality meltdown. Brag about the sweetness of the system sounds, not about the number of doodads available to play with.

FUTURE RECEIVERS

Integration of various audio components has been the trend in the past few years. However, it seems to be reversing, and each component is becoming separate again. I guess people want bulk for their buck. This has its advantages, which were described in Chapter 1.

Dolby Digital receivers are here. They need mass amounts of sound information, but the newer digital audio/visual technology can handle it. As we said in Chapter 4, there are DD-ready receivers that do not contain the digital decoding circuitry but will take the 6 channels. For more information on this, refer to *Receivers and Decoders* in Chapter 4. For more information on the built-in amplifiers, see the previous chapter.

What are a few words to remember when looking for a receiver? Power, controls, Dolby circuitry, not noisy... Oh, yes, and LOOKS! You wouldn't want someone to see you with what looks like a cheap, cheesy receiver, would you?

CHAPTER 7
CASSETTE DECKS, CD PLAYERS, DVD PLAYERS, MINIDISC & PHONOGRAPHS

"I never fully understood the need for a 'live' audience. My music, because of its extreme quietude, would be happiest with a dead one."
Igor Stravinsky

Never in history have we been able to store sounds so accurately and with such a high fidelity as now. In the beginning, there was radio, and the only way to store sounds was on records that weighed as much as a manhole cover. Each newscast was done live. Each radio drama was done with on-the-spot actors. And music was sung or played note by disappearing note (unless they chose to record the music onto expensive phonographs). Then, in 1962, came the ability for anyone to replicate immortal sounds, when the compact cassette was released. A plastic tape coated with magnetically-charged particles held the audio information. Eventually, everyone who owned a hi-fi had the ability to record with the new cassette tapes or play prerecorded masterpieces.

Technology advanced further, but actually took a step back with CD technology. Once again, we were only able to play back what was on a disc (remember the LP?). Times are changing, though. Now we can record onto a MiniDisc and play it back. What's next?

Digital video disc (DVD) is likely to take over the player market soon. DVD-Audio will inevitably be released, and discs only 5 to 8 cm. wide will reveal their secrets to our ears.

The phonograph is, for the most part, dead. Some nostalgia collectors will debate this issue. After all, people still use typewriters and the post

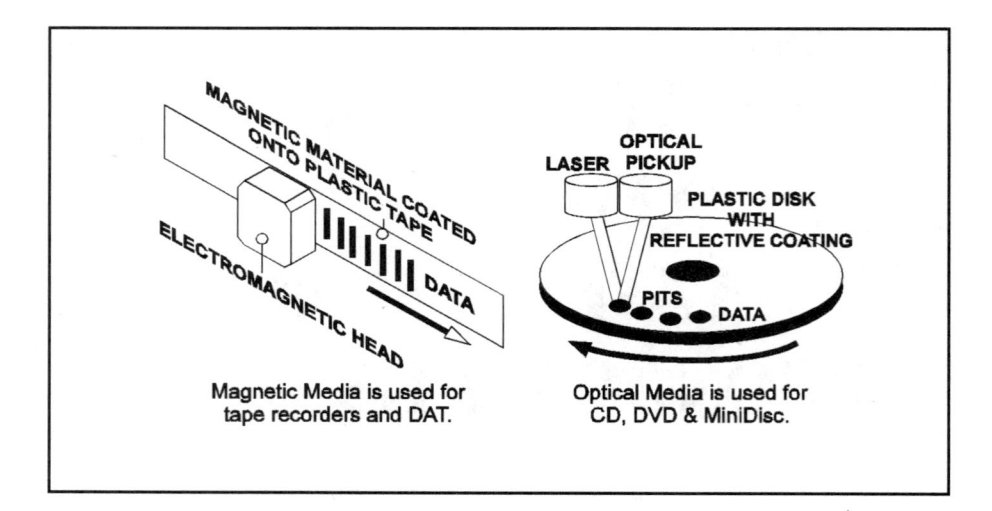

Figure 7-1. Types of recording and playback media.

Magnetic Media is used for tape recorders and DAT.

Optical Media is used for CD, DVD & MiniDisc.

office! Either way, the main focus of this chapter will be on current, popular play-and-record technologies, as well as a hint of future shock, shock, shock... See Figure 7-1.

CASSETTE DECKS

The cassette deck (aside from the high-cost MiniDisc) for the longest time has been the only true piece of consumer audio equipment that can be used to record audio. Some day, an affordable rewritable CD that makes *perfect duplicates* will surface. Right now, similar equipment is reserved only for extremely high-end, expensive audio production, or for the CD-ROM market ("CD burners" are great for storing computer files. Plus, once a writable CD is burned, it cannot be erased and reburned; the initial recording is there permanently). Until then, we have to deal with the tape media and its inherently low signal-to-noise ratio.

The technology to run this piece of equipment is quite extensive because of the racket coming from inside the tape deck. This includes a whirling mechanical and electrical cacophony of noise from the motors, gears and audio heads. Let's take a look at how a tape deck works and how Dolby's magic wand all but eliminates the deck's dissonance.

THE INNARDS OF A TAPE DECK

Like most electronic playback equipment, the cassette deck consists of electrical devices, mechanical devices and sound storage devices, in this case cassette tapes.

Mechanical Devices: These include the motors that drive the cassette's tape around the heads. This part of the tape deck has the potential to create terrible noise in the system. Another annoying factor is that with portable recorders, you are often recording the sound of the motor on top of the intended sounds. Annoying! Other mechanical devices include the loading mechanisms and buttons that engage the heads and motors.

Electrical Devices: These includes the audio heads, the signal amplifiers, the Dolby noise reduction circuitry, the sensors, and various other circuits. A significant amount of noise is created from the tape itself moving across the audio heads. See the Dolby section for more information. The audio heads use electromagnetic principals to encode waves or bits onto a cassette tape.

Cassette Tapes: In the old days of recording, you had to use a reel-to-reel recorder to make audio facsimiles. Then someone came up with the idea of putting the two reels into one small container that is transferable from one machine to another. The tape itself is actually a plastic tape with magnetically-charged particles attached to it. The particles are made of either ferric oxide, chromium dioxide or metal. These materials allow the audio heads to store audio information with the use of an audio-recording head. Many audiophiles still use reel-to-reel, claiming better sound quality. But the very simplicity and portability of the cassette tape ensure its dominance among audio-recording enthusiasts.

HOW THE PARTS INTERACT

The cassette contains the actual plastic tape. When inserted into a deck, a set of audio heads are brought into proximity of the metal-coated tape. The gears and motors drive the tape forward across the heads, allowing information to be recorded or played back.

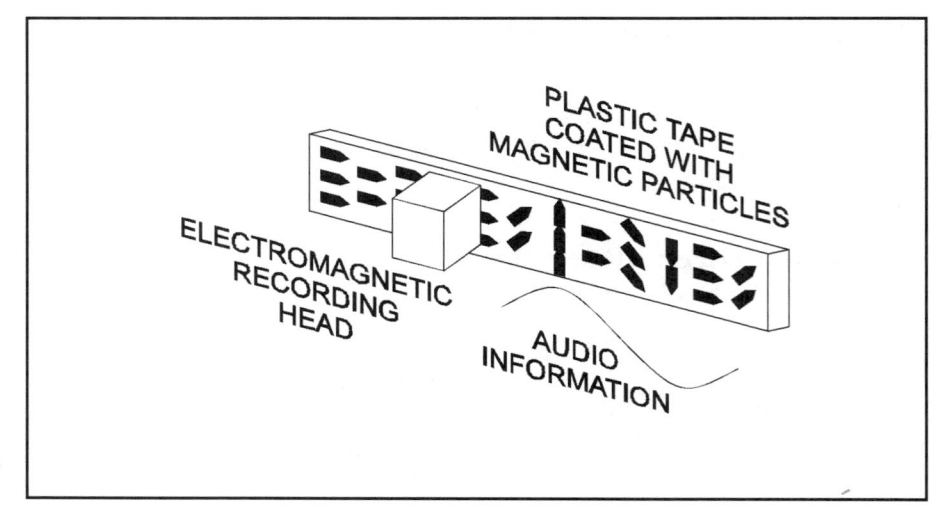

Figure 7-2. Analog recording onto magnetic media.

Recording: The sound signal being recorded is used to energize a tiny electromagnet inside the audio head. It creates an alternating magnetic field that is embedded into the magnetic material on the plastic tape. The magnetic variations cause the tiny metallic particles in the tape to align themselves with the lines of force from the electromagnet. See Figure 7-2.

Playback: After the metallic particles in the tape are aligned, they stay in that position unless the tape is rerecorded. This lets you play back the initial recording over and over. A separate audio head then picks up the faint magnetic signature embedded into the tape during recording; the coil inside the head reacts to the magnetic particle's positions. This signal is then amplified to usable levels and sent to the receiver or power amplifier. Out comes a sound signal for the speakers.

TYPES OF TAPE DECKS

There are two basic kinds of tape decks: a single unit and dual deck. See Photo 7-1. The deuce model dominates because you can use it for dubbing or to carousel two tapes. The other differences between cassette decks have to do with the tapes they can play and the noise elimination circuitry they use.

> **Dolby Saves the Day!**
>
> Dolby Laboratories designed a technology that significantly reduces the biggest culprit of distortion and noise in tape recordings. I am talking about the noise from the tape moving over the heads. Besides this, Dolby technology improves the signal-to-noise ratio that comes from the narrow recording track and slow speed of the tape moving across the head. If you have ever turned up the volume and listened to a silent passage on a cassette tape, you get an idea of what Dolby Noise Reduction (DNR) does for analog tape recordings. If there is no DNR, you will hear a hissing and popping sound. The silent passage is supposed to reflect *silence*, not sound like a snake! Hit the DNR button, and the hissing will then sound more like the original recording — quiet.

DIGITAL AUDIO TAPE: DAT

It is surprising that this high-tech digital cassette has not become a smash hit in North America like its digital disc descendant, the compact disc (CD). A DAT can create copy after perfect copy of an original CD recording without degradation. It is basically an audio "Xerox machine."

Its workings are very similar to a CD, except it uses a special cassette tape for storing the digital data. The matchbox cassette holds up to two hours of accurate audio. The problem is that the players are around three to four times the cost of their analog cousins, the cassette decks. But if you are into high-end audio recording, mixing or dubbing, it is currently the best way to go.

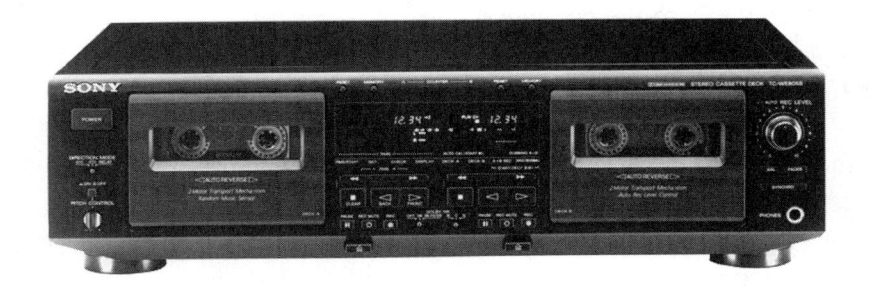

Photo 7-1. Dual-deck stereo cassette player. Reproduced with the permission of Sony of Canada.

Chapter 7: Cassette Decks, CD Players, DVD Players ...

Its cost is still very high because there is a surtax on DAT merchandise. Why? Because it is the music industry's worst nightmare! Giving millions of people the ability to make a perfect facsimile of their albums over and over? Absurd! Who do you think gets the DAT surtax payments? You guessed it: the big record labels.

NOTE: Is DAT worth the investment? If you want a high-quality recording device, yes! But then again, the MiniDisc and other recordable device prices are falling faster than the record company's profits. Record-once CDs are down to almost a buck each, and the burners are becoming affordable at $250-500. The rewritables are likely to match these levels with time.

WHAT TO LOOK FOR IN A TAPE DECK

If you don't want to pay the premium for the new DAT decks, try to get a good dual-deck with DNR, microphone inputs, adjustable bias control, auto rewind and CD synchro. A tape counter is also nice, and make sure the deck provides *low flutter* or *wow*.

COMPACT DISC PLAYERS

CDs are probably the most popular consumer electronics device of all time. The use of a 5" plastic disc for storing bits of information was the audio and computer market's savior. These bits can represent letters, numbers, pictures or (in the case of audio) sound waves. Audio waves are analog; there is nothing digital about them. So, they must first be converted into something a digital device can understand, then converted back into analog so our stereo systems will comprehend them. Let's first take a look at how these round music mirrors work, then go into what's available to the consumer.

LET'S GET DIGITAL

Refer to Figure 7-3. The compact disc itself has pits and lands, with a laser beam aimed at them from the CD player. A land reflects most of the laser light back to a device called a photocell. A pit does not reflect, or the reflection is weaker. The photocell is an electronic component

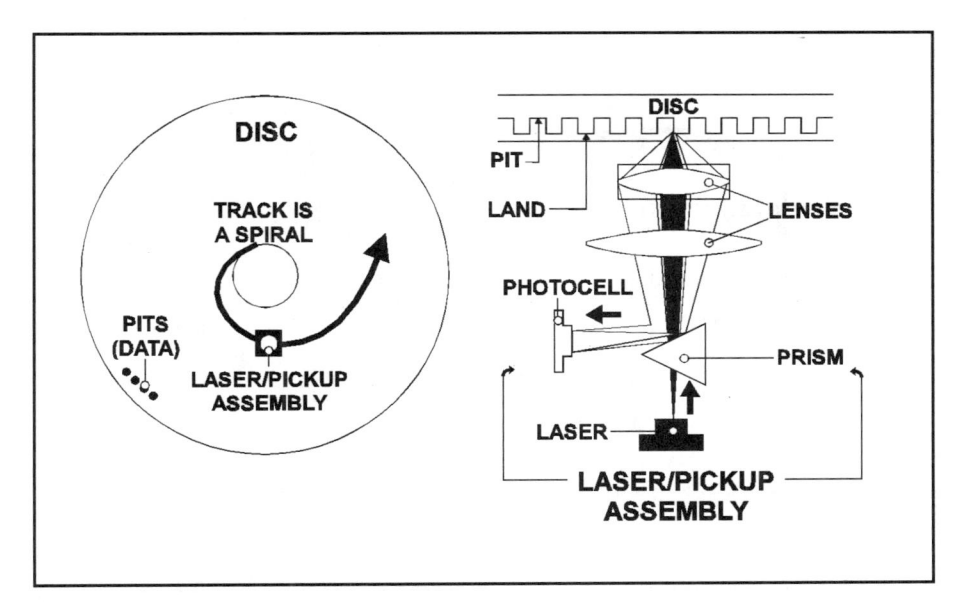

Figure 7-3. Compact disc mechanics.

that changes the laser light into electricity. This is used to determine if the space on the CD is either a one or zero (bear with me, I will explain the ones and zeros soon). So, if the laser hits a land, the light is reflected to the photocell, which then tells your CD player it sees a "one." The pit tells it there is a "zero." The ones and zeros are then converted into an analog sound wave.

NOTE: The actual method that a CD uses to store information with the use of pits and lands is much more complex than stated. However, for sanity's sake, keep in mind that a pit equals 0 and a land equals 1.

WHAT ARE THE ONES AND ZEROS?

"Digital" implies the use of the binary number system. There are only two numbers in this system — 0 and 1. These numbers actually represent "on" and "off" commands. We use this number system in electronics because transistors and microprocessors only have two states, on and off. Just like a light switch.

Each switch represents one bit. If we string eight of these bits together, we can represent any decimal number between 0 and 255. This is called a *byte*. See Figure 7-4. The more bits we add, the larger the number we

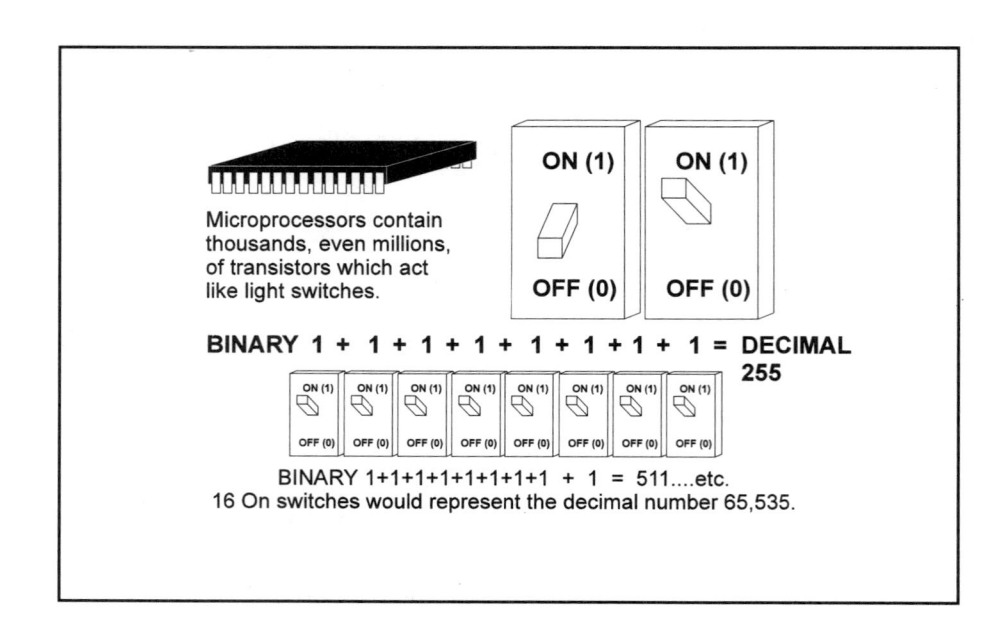

Figure 7-4. Digital explained.

can represent. Actually, one bit doubles the number we can represent. So, one bit added to an eight-bit number can represent decimal 0 to 511.

So, when you hear the word *digital*, it simply means that the equipment uses the binary number system. When you hear the term *16 bit* in reference to a CD player, it means that each section of the sound wave is stored with 16 bits. (More on this later.)

ANALOG-TO-DIGITAL & DIGITAL-TO-ANALOG CONVERTERS

This is the whole basis behind a CD player. Why do you need to learn about all this digital information? Because a CD uses these bits and bytes to create a sound wave. It does this through an electronic device called a digital-to-analog converter (DAC). The DAC takes the digital information being fed to it by the CD and converts it into an analog wave. Analog-to-digital (ADC) devices are simply the reverse. They are used to turn the analog sound wave into numbers that can be stored digitally on a disc.

DACs are classified by how many bits they use to convert a binary signal to an analog signal. If the DAC has 256 different voltage levels, then it has a resolution of 8 bits. Most CD players have a 16-bit DAC,

NOTE: A DAC outputs a given voltage level for the binary number being fed to it. If you spread this level out over time, you create a wave.

which has about 65,536 different voltages. The higher the number of bits, the greater the accuracy on playback, and the better the signal-to-noise ratio.

HOW A CD CHANGES DIGITAL INFORMATION INTO AUDIO WAVES

Refer to Figure 7-5. When the music was originally recorded, the ADC took a voltage measurement of the input audio wave every split second. This is called *sampling*. In the next split second, it took another voltage reading. The machine stored each split-second 16-bit reading onto the CD as a binary number between 0 and 65,535.

When the CD is in your machine and the laser is reading the numbers, the DAC can turn them back into their original analog voltage values. A number that has been read from the disc is fed to the DAC, usually a 16-bit number. This sets the DAC to a certain voltage. A fraction of a second later, another number is fed to the DAC and out comes the voltage. On and on the process goes until something that resembles the original analog sound is sent out to your speakers.

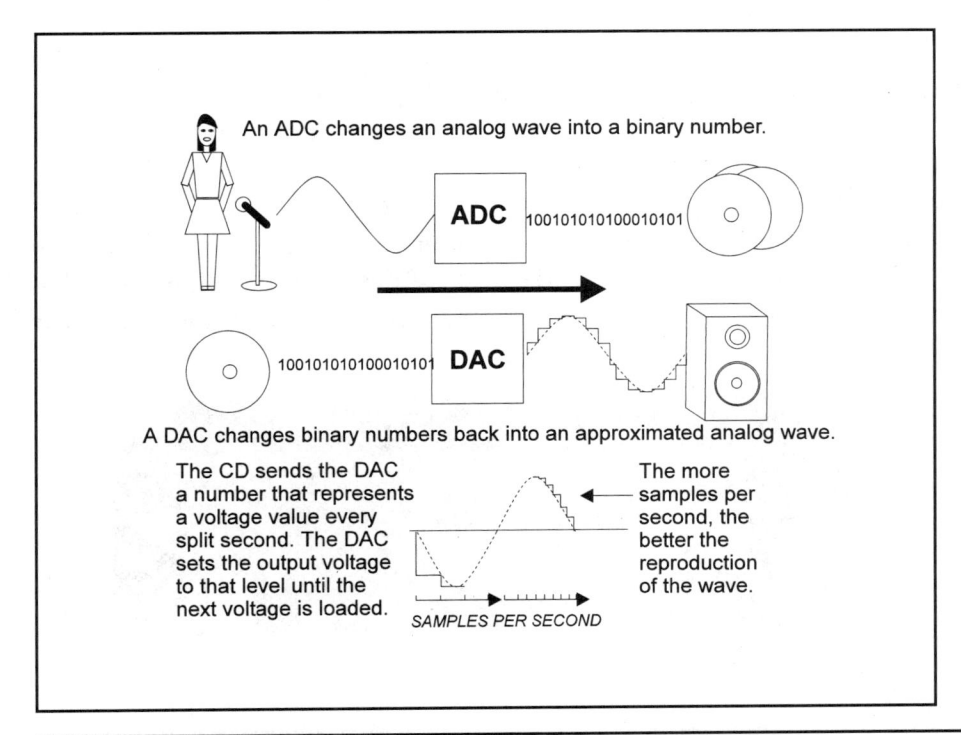

Figure 7-5. ADC and DAC.

SAMPLING

As we said, sampling is taking measurements of the input sound signal at a fixed time interval. A CD needs to know the amplitude and the polarity of the wave (if it is positive or negative). The player does this at a lightning-quick pace. In fact, the process is twice the frequency of human hearing: 44,100 times a second. Because each wave has a positive and a negative half, the maximum number of complete waves the CD measures per second is 22,500.

OVERSAMPLING

Converting a digital signal into an analog signal is called *quantization*. In doing so, the DAC produces unwanted quantization noise that needs to be filtered out before it hits your speakers. This is done using oversampling. It takes the sampling into an inaudible range so it doesn't find its way into your speakers. It virtually eliminates the side effects of quantization (ringing in your ears).

CD PLAYERS

There are two basic varieties of players. Traditionally there is the single player. These are found on lower-end CD players. The newer category, and the most popular by far, is the CD changer, or (in some cases) the carousel model. These jukebox emulators can hold from 5 to 200 CDs, all available at the push of a button. See Photos 7-2 and 7-3. Either CD player will play near-perfect quality sound. The numbers and features (such as 1-bit sampling, MASH, and 8-times oversampling) are relatively unimportant, as you will not likely be able to hear the difference

Photo 7-2. 5-disc carousel CD changer. Reproduced with the permission of Sony of Canada.

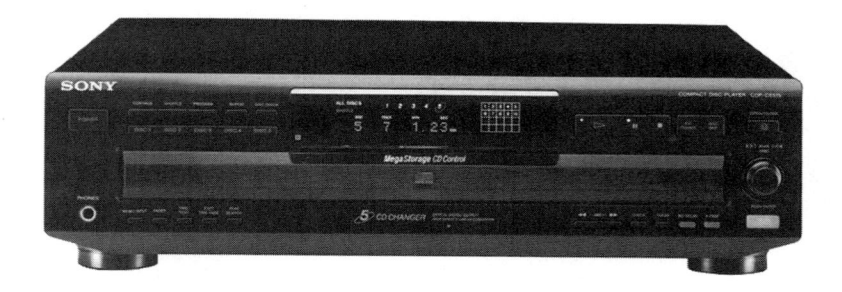

The Great Digital Debate

Audio purists still think the CD player should never have been invented. They argue that the sound is too plastic-sounding, harsh, mechanical, fake, bland, or too "CD-ish." No character! Of course, they may not realize the playback is nearly a perfect version of the original recording. In other words, the CD player does not add a lot of distortion to a recording the way a tape or LP would. This is irritating to people who want the audio equipment itself to add a persona and depth to their favorite LPs. For the most part, the argument has all but died, and CD sound is globally accepted save for those holding out by clutching to their vinyl memories. But then again, that Van Halen album sure sounded better on my turntable.

One other point to remember is that vinyl LPs require intense maintenance and care so they do not break or warp. CDs, on the other hand, can almost survive a nuclear holocaust (or young child with candy) and still come out with no degradation in sound quality. DVDs are even more baby food-proof.

Photo 7-3. 200-disc megastorage CD changer. Reproduced with the permission of Sony of Canada.

between these higher-feature, higher-cost units and a vanilla-flavored model. It is more important to find a unit with the features you specifically need than to worry about how the "bit stream" is delivering sound to you.

Chapter 7: Cassette Decks, CD Players, DVD Players ...

Photo 7-4. DVD player. Reproduced with the permission of Lenbrook Industries, Marantz Canada.

DIGITAL VIDEO (VERSATILE) DISC PLAYER

We have described the features and fantasies of the DVD throughout this book. I only wish to add a few comments.

Because a DVD player will play regular CDs in addition to video discs, you may want to purchase one in lieu of a CD player. See Photo 7-4. Or you may want to wait until the new DVD-Audio machines are out. If you want a serious home theater, try to get a DVD now. The picture and sound quality is so incredible, I can only compare it to the difference between mono and stereo. It makes you wonder what's next. The future of DVD is recordable. If this doesn't evolve for DVD, another device along these lines will. Soon.

MINIDISC: MD

The MiniDisc is the only re-recordable disc device available, except for the emerging rewritable drives for computers. See Photo 7-5. It has

---

Should You Buy a DVD Player Now or Stick With a CD Player?

This has been a common question since DVD players came on the scene. Here is how to decide: yes, the DVD player will play regular CDs, but it is currently aimed at home theater use only. If you are a serious home theater player and want a top-quality signal for watching your favorite movies, go for it. If you simply want it to play CDs, wait until the DVD-Audio is released. Otherwise you are overspending up to 300 to 400 percent, and are likely to have to buy a new unit when the technology changes. Invest in a quality CD changer and more CDs instead. At least you can enjoy them NOW!

Photo 7-5. MiniDisc recorder and player. Reproduced with the permission of Denon Electronics.

been around since 1992, but has not risen in popularity until recently. There are two varieties of these discs, prerecorded and recordable. The recordable versions are highly popular. CD still reigns as the king of the prerecorded market, and no one wants to have to spend $15 for a CD and $15 for a MiniDisc of the same album.

THE DISCS

The MiniDisc is a 64-mm. diameter disc that stores audio information. The disc is inside a tiny 2½" square cartridge that protects its coatings. The prerecorded disc is the same technology as the CD, and stores up to 74 minutes of recorded material. The recordable discs use a technology called *magneto-optical*, or MO. It lets you record and re-record audio up to *one million times* and is claimed to last 30 years or more!

ATRAC

Adaptive transform acoustic coding (ATRAC) is an ingenious audio compression technique that can crunch the data from 5 minutes of audio into 1 minute of space. This is why the discs are so small (less space needed to store the same amount of music). It works by trashing all the audio information that we are not likely to hear or is being masked by dominant sounds. The problem with this sound squeeze is that the model used by the MD to determine what we hear and can't hear is not perfect. Therefore, the engineers are continually perfecting the model. This is what *ATRAC 2* or *ATRAC 4.5* means. The higher the number, the higher the quality of the sound.

NOTE: There is a subtle difference between the quality of an MD and a CD because of the ATRAC compression scheme. However, most people won't be able to detect the slightly degraded quality in MD devices.

HINT: The higher the ATRAC number, the higher the cost, and greater the quality of sound of a MiniDisc device.

Chapter 7: Cassette Decks, CD Players, DVD Players ... 83

MINIDISC RECORDERS

Because MiniDisc is not as popular as CD technology yet, costs can be high. However, in the past year or so, decks have gone down in price exponentially. Most manufacturers produce home decks, portables, car stereo and boom box models for $150 to $700. A respectable deck for your home sound system is down to $200 or so. Well worth it! Prerecorded MiniDiscs are about the same price as CDs, and blank cartridges are about $5 a shot, currently.

THE PHONOGRAPH

Since the CD has become such an icon of audio society, the phonograph has been spun back to the middle ages in most minds. Don't always believe what the digital press has to say, though. LPs add ambiance and character to a recording. A CD is simply an exact low-distortion copy. Turntable equipment choice and setup determines the character of the sound. With CDs, you are at the mercy of perfection. Terrible, at least to some audiophiles. Records are still an audio market, and you should leave your mind open to this venerable piece of equipment.

PARTS OF A RECORD PLAYER

There are three components to a typical record player. The turntable is the device that revs your records. Attached to that is the tone arm, which holds the third part: the cartridge. The turntable and tone arm are usually a package deal, but swapping a cartridge is possible.

HOW AN LP WORKS

Sounds can be recorded onto a vinyl disc (LP). An analog wave that contains music is engraved onto a continuously-spiraling track. When the LP is placed on the turntable, the tone arm moves over to the periphery of the disc and drops the cartridge's needle into the groove. The needle is then moved back and forth ever so slightly by the record's grooves. The to-and-fro movement of the needle is transferred to either an electromagnetic or piezoelectric device that changes the mechanical motion into electric signals. The low-power signals are then sent to your preamplifier's phonograph input, which then amplifies the signals to usable levels.

PHONOGRAPH RECOMMENDATIONS

It's funny, but true: some hi-end hi-fi equipment still has LP record players attached when purchased, but not CD players. Most hi-end audio companies still make turntables and cartridges. However, the cost of quality equipment is out the roof. If you can, get an old turntable and order a quality cartridge from Shure, Denon or Audio Technica. Play with different setups and, most importantly, take good care of your LPs, making sure to clean and store them properly. Remember, they aren't made anymore!

WRAP UP

The market has basically made the choice regarding how you listen to new prerecorded music. CDs are the popular medium, so CDs are cheap and abundant. Cassette tapes are still around, but disappearing just as fast as LPs. I know there are still vinyl maniacs that will violently disagree, but the reality is that you just can't buy new music on LPs. There is, however, the choice between a CD or MiniDisc system and DVD. The MiniDisc is available *now*, and is actually an amazing consumer audio product — just not popular. For now, try to stick with a quality CD changer and dual-cassette deck. If you already have a turntable, treat it well. If you want to do any type of recording, consider the MiniDisc or DAT. DVD-Audio has not yet arrived; but when it does, it will likely be the successor of the CD player. I am sure in 20 more years, our kids will be asking, "The 'compact' what?"

CHAPTER 8
LOUDSPEAKERS, HEADPHONES & MICROPHONES

A device that converts energy from one form into another is called a *transducer*. Speakers, microphones and headphones are examples of *transducers*. They turn sound waves into electrical waves, and electrical waves into sound waves. Microphones change acoustical energy into electrical energy. Speakers and headphones change electrical impulses into physical movement, much like a car engine converts gasoline into motion. Our ears interpret this alternating acoustic energy as

Thack shows off his new stereo.

Chapter 8: Loudspeakers, Headphones & Microphones

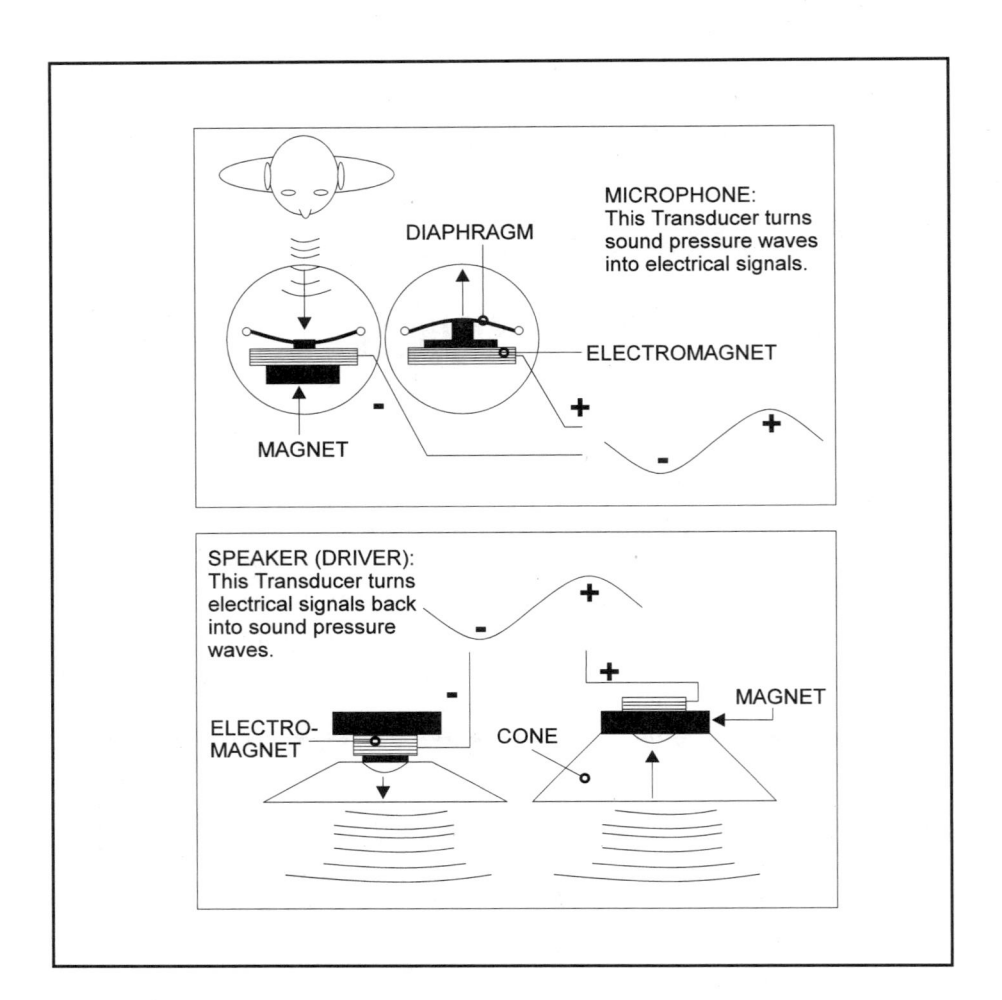

Figure 8-1. Sound transducers.

sound. The most common transducer used in audio is one that utilizes electrodynamic principles, using electromagnets and permanent magnets. See Figure 8-1.

LOUDSPEAKERS

Loudspeakers are the stereo system's most important components. See Photo 8-1. They determine the final outcome of your audio investment. This is the output stage where the amplified audio signal is converted back to acoustic energy. Because the speakers are the end of the line for the electronics, and the beginning for the actual sound waves, it is important that the loudspeakers are well constructed, physically and electronically. Speakers make or break a well-packaged stereo system.

GETTING IT RIGHT

Loudspeakers are typically a stereo system's weakest link. Bad designs are prone to clipping and distortion. Haphazardly-placed loudspeaker cabinets negate the stereophonic effect the designers worked so hard to achieve. Mismatched impedance between the drivers and the amplifier can cause either the speakers or the amp irreversible (expensive) damage. The solution to these problems is knowledge. Once you know the workings of a loudspeaker and how to improve its performance, you will know how to maximize the sounds (and get the most for your audio investment).

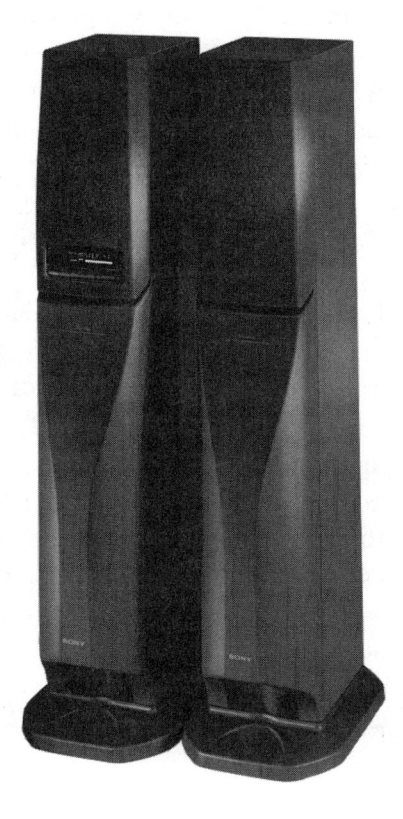

Photo 8-1. Home theater speakers with built-in decoders for Dolby Pro Logic. Reproduced with the permission of Sony of Canada.

DRIVERS

The actual speakers in a loudspeaker setup are called *drivers*. They are the transducers. Most speakers contain three types of drivers, each operating on the same principle of cone and electromagnet. The highly-amplified audio signal is sent to the driver's electromagnet, which reacts through a permanent magnet. It creates a pushing and pulling action that reacts with the attached cone. This displaces air, just like in a piston: back and forth, back and forth at different frequencies and amplitudes, to recreate the original sound waves that were fed to a microphone at some point in time.

NOTE: Most drivers are of the electrodynamic variety. This means they use a permanent magnet and electromagnet to pump a cone back and forth. Flat electrostatic speakers are available, and may someday replace dynamic speakers. Don't hold your breath, though.

Chapter 8: Loudspeakers, Headphones & Microphones

COMPOSITION OF A LOUDSPEAKER

A modern loudspeaker is divided into three basic electronic/mechanical components: a cabinet, drivers and a crossover network. Refer to Figure 8-2.

CABINET OR ENCLOSURES

Loudspeaker cabinets are made of a variety of materials. The type of material used somewhat affects the acoustic properties: the stiffer the material, the less quirks in the acoustics; the lighter the material, the better your back will feel after lugging the speakers around. An average loudspeaker is composed of medium density fiberboard (particle board) with veneer or vinyl finishes. It is cheap and relatively light (compared to some speakers made of marble or formed concrete. Yes, concrete!). The dimensions of the cabinet are carefully calculated to work with the drivers and crossover network, so it is best to stick with store-bought loudspeakers. The engineering is included in the price.

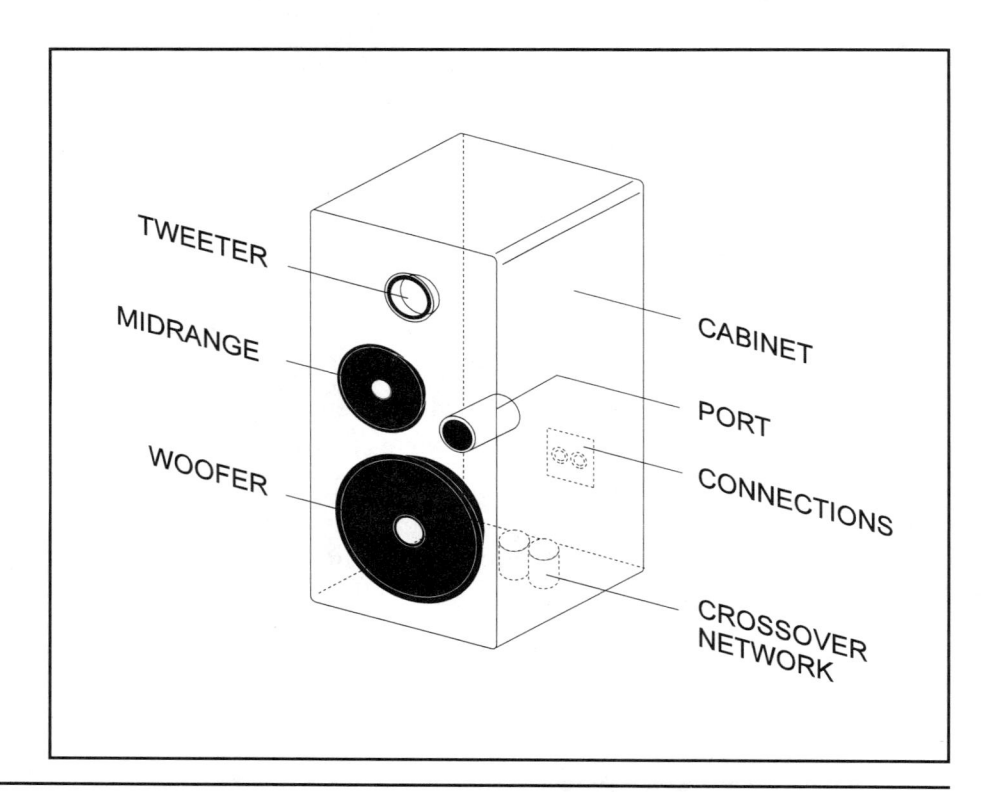

Figure 8-2. Typical loudspeaker.

The speaker cabinet has a dual job. One is to hold all the speaker components, and the other is to prevent the driver's rear wave from causing destructive interference with the front wave. Remember that the driver has a back and front wave, which are exactly opposite. See Figure 8-3. If the two mix, they cancel each other out. There are two types of cabinets that solve this problem:

Acoustic Suspension: This type of cabinet is completely sealed, preventing any sound from radiating from the rear of the drivers. By blocking in the rear wave of the driver, it can never escape its cage and destroy its creator.

Bass Reflex: The enclosure allows the rear wave created by the back of the drivers to bounce around the inside of the cabinet and then be radiated out into the room. Usually, a small opening called a *port* is placed somewhere in the cabinet's surface. This puts the rear wave in sync (phase) with the front wave, then lets it out the hole.

DRIVER TYPES

The driver is the actual speaker in the loudspeaker cabinet. There are usually three types of drivers built into the cabinet (unless it is a sat/sub

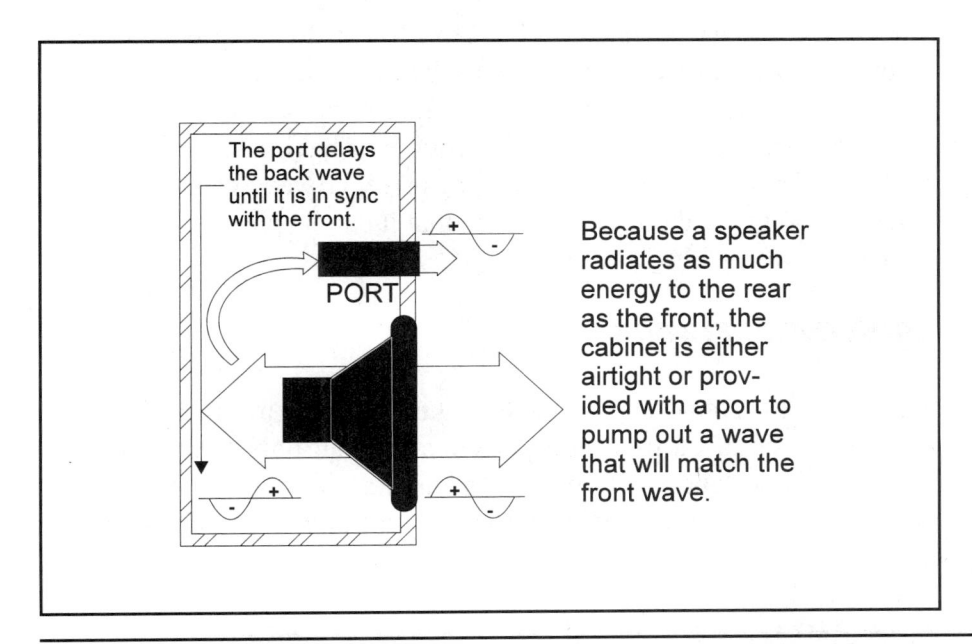

Figure 8-3. Bass reflex.

Chapter 8: Loudspeakers, Headphones & Microphones

combo, described later). Following are descriptions of a subwoofer's drivers, which are typically in a separate cabinet. Each driver plays a certain range of frequencies and has its own characteristics:

Tweeter: High-frequency driver. It is the smallest driver, and usually looks like a shiny dome. Tweeters are very directional and sensitive to overdriving/overheating. Typically, the tweeter is filled with a type of heat-dissipating fluid to combat this problem. It produces frequencies that travel a very straight line, which is why an inverted dome is used — it sends out the signal in all directions. Range is usually 4 kHz to 20 kHz. The tweeter can be as small as one inch in diameter.

Midrange: Medium frequency driver. This is a medium-size cone-type driver, which usually provides the voice frequencies. Together with the tweeter, midrange drivers create the stereo image. Range is usually 800 Hz to 10 kHz, and size is 3 to 8 inches in diameter.

Woofer: Low to midrange frequency driver. Woofers are typically the largest cone inside the enclosure. This is the bass driver. A woofer spits out frequencies that can bend around objects with ease. Thus, it is not important to have them pointing directly at you. Range is usually 20 Hz to 2000 Hz, and size is 10 to 20 inches diameter.

Subwoofer: Low frequencies, 20 Hz to 150 Hz, do not affect the stereo imaging, so a separate cabinet with a subwoofer can be connected to the amp to do all the thumping! It doesn't need to be part of the two stereo speakers. This saves space and the trouble of having to make a larger enclosure for the smaller stereo imaging drivers (tweeter and midrange). What is also great about this driver is that it can be placed anywhere in the room. Even within a wall.

CROSSOVER NETWORK

The amplified sound signal needs to know where to go, because the various drivers used in a loudspeaker pump out only their respective tones. This is done electronically through a crossover network. The crossover takes the original input signal and divides it into frequency ranges. These spectrums are then sent out to the proper driver. See Figure 8-4.

HOW A LOUDSPEAKER WORKS

A signal from the CD player, tape deck, tuner, etc., is sent to the power amplifier. A huge amount of power is added to the signal and fed to the crossover network, which separates the frequencies for their appropriate driver. The amplified signal is then used to power the driver's electromagnetic coil with either a positive or negative side of the wave. If it is positive, the magnetic coil is thrust forward by the opposing magnetic fields from the permanent magnet. The attached cone is consequently forced forward along with the surrounding air, making a high-pressure area. When the wave goes into its negative segment, the cone is pulled in and creates a rarefaction of the air. The two pressure waves are then sent out for our ears to hear.

COMMON LOUDSPEAKERS

There are basically three speaker configurations for both stereo hi-fi and surround sound. Take a close look at the pros and cons of each to help you decide which ones you'd like to use. See Figure 8-5.

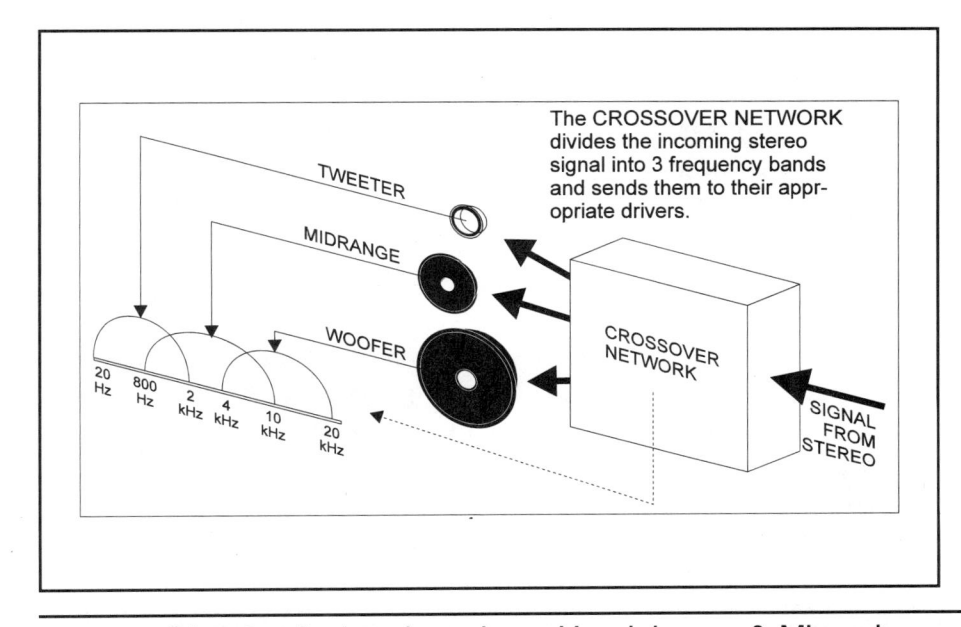

Figure 8-4. Crossover network.

Chapter 8: Loudspeakers, Headphones & Microphones

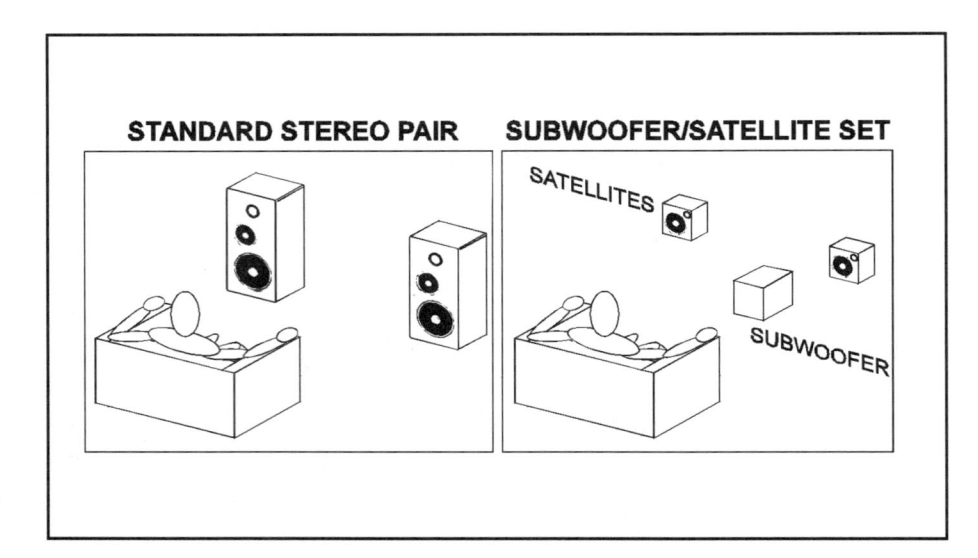

Figure 8-5. Speaker configurations.

CONVENTIONAL FRONT SET

These are used for a standard stereo pair setup, consisting of two loudspeakers, each in a fairly large tower-type cabinet. They're fine for normal stereo listening, but can't recreate a surround sound image on their own.

Pros: They are a one-box package deal with a set of three drivers in each. Some people say they actually sound richer than a sub/sat setup. Decide for yourself by listening to both in a store.

Cons: The enclosures are heavy and unnecessarily bulky to accommodate the larger woofers. This results in wasted space where the tweeters and midrange sit.

SUB/SAT

HINT: If you are building a surround sound system, you can use four satellites, a medium-size center speaker and a subwoofer.

Sub/sat means *subwoofer and satellite combination*; also called a three-piece system. Sub/sats are always used in conjunction with each other. Speaker designers came up with an elegant solution to tackle the bulk problem of conventional speakers. They placed the tweeters and midranges in two small cabinets, which lets you hide them on a tight shelf and listen to the mid- to high-frequency sections of music. The

94   Howard W. Sams & Company **Complete Guide to Audio**

single, larger subwoofer is enclosed is a gargantuan box and placed out of sight. It produces that chest-pounding bass effect.

Pros: This setup allows for a very inconspicuous entertainment center. If you live in close quarters, all you have to do is place the two small satellite enclosures on a shelf and the subwoofer in a corner.

Cons: Some people say they are not as robust as conventional loudspeakers; but then again, some people mentally relate the size of the speaker to the volume or sound quality.

SATELLITES

These micro machines punch out a whole lot of power for their Lilliputian size. Satellite speakers typically contain a midrange and a tweeter. Several companies have put most of their research and development into satellites in order to maximize size and make them interact with the subwoofer without annoying distortion problems. BOSE, Cambridge SoundWorks and JBL have particular quality product, but most manufacturers put out a sub/sat set.

SUBWOOFERS

Subwoofers come in different shapes and sizes, but the typical unit is a wood or plastic box from the size of a computer case up to the size of small end table. See Photo 8-2. Inside sits a 10- to 18-inch diameter driver that can pound the surrounding air. They operate from about 10 Hz, and top out at around 200 Hz; 25 to 80 Hz being their normal range. Most will deliver up to 110 dB will little distortion or damage.

NOTE: Satellites remind me of the Stealth fighter jet: invisible, but loud! They nearly disappear into your environment, but continue to give out quality sound from seemingly no particular direction, if set up correctly.

Photo 8-2. Subwoofer with a built-in amplifier with 30 watts of power. Reproduced with the permission of Sony of Canada.

Chapter 8: Loudspeakers, Headphones & Microphones

> **Subwoofers**
>
> Subwoofers are the new got-to-have-it item of the home theater revolution. With Dolby Digital surround sound, a subwoofer brings a realism to movie sound effects that were previously available only in a theater. Now you can bring home those pounding T-rex footsteps from *Jurassic Park*, and that freight-train-thunder of the tornado from *Twister*.
>
> There are two types of subwoofers. The older model is called *passive*. This means that the amplification is coming from an external source such as a receiver's amplifier. The newer, more popular active subwoofers have built-in electronics and amplification. This prevents you from having to buy an overpowered A/V receiver and running heavy wire to handle the flood of power used by the driver. Simply place the subwoofer by a power outlet and run a small signal wire from the receiver to the subwoofer's innards.

Subwoofers are nondirectional; or more accurately, omnidirectional. You can place them anywhere in a room, preferably as close to the receiver as possible so you don't have to run as much wire. By using a subwoofer with a built-in amp (active), you can prevent wiring woes.

SURROUND SOUND SPEAKERS

A surround sound setup depends on the type of Dolby system you are using. A standard A/V speaker package typically consists of a front stereo pair, a center dialog speaker, two surround sound speakers, and sometimes a subwoofer. You can start with the two front speakers that came with your stereo, and add a specially-shielded center speaker. Surround sound is described in detail in Chapter 4.

HOW TO SET UP YOUR SPEAKERS

NOTE: You can forego the floor plan if it is simpler for you to plan with a long tape measure and a helper. Penciling out your plan saves time and energy.

Pay attention. This is the most important datum in audio equipment: *Set up your speakers correctly*!

Now that this is burned into your brain, and you are having nightmares about accidently moving a speaker sideways one foot, let's see how to properly set them up. See Figure 8-6.

Which Speaker Setup Should You Choose?

Should you go for traditional twin towers, a space-saving sub/sat set, or surround sound?

First, ask yourself two questions first:

"What do I have space for?": For most people, the answer can be, "All of the above." But do you want a bunch of large monoliths planted in your living room, or would you rather have the technology go unnoticed? Some people absolutely want the largest possible equipment for the impression it makes, thinking BIG equals POWER! For speakers, this is not true anymore.

"Which sounds better to me?": If you consider yourself an average person with average sound tastes, you will not likely hear a difference between a twin pair and a sub/sat setup. However, high-end audiophiles seem to share an opinion that the sub/sat combo lacks control over the entire frequency band. I tend to believe this after listening to many screechy satellites and underpowered subwoofers. You may be able to find an appropriate combination, but perhaps the decision should ultimately be left to your own sound preferences.

1. Draw a dimensionally-accurate floor plan of the room that will contain audio equipment.
2. Roughly determine the layout of each piece of equipment (rack, speakers, TV, etc.) and the furniture placement. Pay close attention to where you will be sitting.
3. Place an "X" at the approximate place your head will be in the place where you plan to sit. Draw a straight line from the X to the wall in front of you, where the speakers will be set. This will be your *center line*.
4. Take a protractor and draw a line from the X to the front wall, 30 degrees from the center line. Repeat this on the other side of the center line. Where these lines meet the wall are where the speakers should be placed. Try to make sure that the distance between the speakers is between 6 to 10 feet, and forms an equilateral triangle with your seat. If not, move the seat and redo the last two steps.
5. Make sure nothing is blocking the path of sound from you to the speakers.

Chapter 8: Loudspeakers, Headphones & Microphones

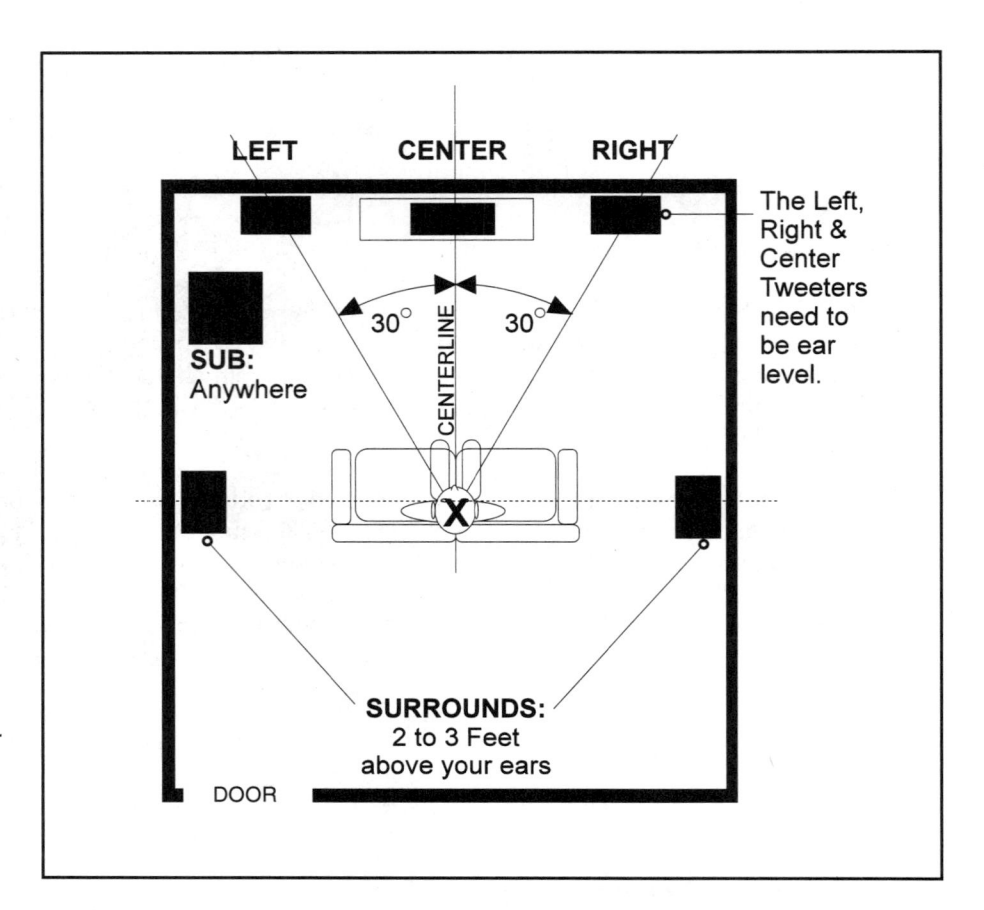

Figure 8-6.
Setting up your
speakers.

6. Make sure the height of the tweeter matches the height of your ear. Remember that this is the most directional driver in the speaker. The straighter the line to your ears, the better it will sound.
7. Listen to some music, especially a well-done stereophonic piece, and see if you can determine the direction of each instrument or voice. If not, move both speakers 6 inches *toward* each other and listen. Did the sound quality get worse? If so, then move the speakers one foot away from each other. Fine-tune the distance until it sounds as if you can just reach out and touch the musicians and their instruments.

If you cannot get the stereo effect despite fine-tuning, then the polarity of the speaker wiring may be reversed. Make sure the red wire (or raised bead side) goes to the "+," or the red connector on both the speaker and amp. Make sure the black wire goes to "-," or the black connector.

This will make both speakers push out and pull in at the same time. Otherwise, they will cancel each other out.

SETTING UP SURROUND SOUND SPEAKERS

1. Set up the front speakers as described in the last section.
2. Place the center speaker directly in line with the center line you previously drew. The height of the speaker should also be the same height as your head (usually placed on top of the television).
3. Set the surround speakers to the sides of your head, 2 to 3 feet higher than your ears. Do not move them behind your head, or the surround effect is lost. You can wall mount these or use stands if the room is too large (like anyone has too big a room!).
4. The subwoofer can be placed anywhere in the room. Try to get it as close to the receiver as possible so you won't have to run a long cord.
5. Systems with Dolby Pro Logic have a test signal generator that lets you test the balance of all the channels. Engage this feature (check your manual). As the signal travels from channel to channel, adjust the balance controls on the receiver until each channel plays at the same loudness level.

POWER RATINGS

Most speakers have a power rating printed on them somewhere. Try to match this, as well as the impedance level with your receiver, as best as possible. See Chapters 5 and 6 for more information on impedance and wattage ratings.

DIGITAL SPEAKER SYSTEMS

A new type of speaker is likely to take over the audio market as more and more digital equipment comes into use. It is called the *digital speaker system* (DSS). Instead using the receiver to send an amplified signal to your speakers, it sends a digital signal. Inside each speaker is a *digital signal processor* (DSP) and built-in amps. This prevents distortion caused by lengthy speaker wiring. Also, the speakers can be hooked directly to a CD's digital output without having to go through a receiver.

This also cuts down on noise. A special DSS link can chain up to 12 DSS speakers off a single device. Keep an eye out for this new technology. It may mean replacing every component once again. Ugh!

HEADPHONES

Headphones are used all the time with a Walkman or other portable tape player or CD player. But what about those times when your home stereo is just a little loud for others? Walkman-type headphones with their spongy ear pieces may not be your best option. A pair that encases your ears, minimizes external noise, and immerses your auditory senses in the musicians' world may be the answer.

WHY HEADPHONES SOUND DIFFERENT FROM SPEAKERS

When you listen to sound coming from speakers, or even live sound, your head and body actually influence the treble range, causing an apparent boost of those frequencies. This is called *acoustic diffraction*. With headphones, the sound is sent directly to your ears and the diffraction is nonexistent. So, if you listen through headphones to the same signal being sent to the speakers, it will sound as though the treble needs to be jacked up. However, a good set of headphones will compensate for this by hiking the treble up slightly to emulate the same sound from your speakers.

SEALS

The pads that encase your ears are called *seals*. Stereo headphones can be categorized according to the acoustical properties of these seals:

Open Seals: These let in a high amount of surround sounds, letting your sense of hearing remain partly in the real world.

Semi-Open or Semi-Closed Seals: These let you listen to the music, but also allow you to hear slight external noises such as a kettle whistling or someone calling for your name.

Closed Seals: These cut you off from real-world sounds by allowing very little sound leakage to reach your ears. This allows you to transport your mind into the music.

Try not to used closed-seal headphones. You may be cutting your ears off to emergency sounds, such as a smoke detector. You can buy semi or open seals and always say, "But I didn't *hear* you asking me to take out the garbage."

WHAT TO LOOK FOR

Each set of headphones has its particular sound. The kind to look for is one that reproduces the low- to mid-bass frequencies well, as these are rarely modified with the tone control. If you are going to be sitting quite a distance from the stereo, either make sure the cord is long enough or get a cordless model.

Cordless headphones are wonderful, but be sure to test listen before buying, as some models are more prone to distortion and interference than others. Also, there are two different types of cordless headphones: digital and radio. Digital headphones are much more expensive than radio, but are worth it due to the sound quality and the range. Many digital headphones have a 60+ foot listening range, allowing you to walk around your neighborhood block or office building without losing the signal. Most cordless headphones are rechargeable. They come with a recharging base and transmitter. They also come with connections that allow you to listen to whatever system you like, from stereo hi-fi to Walkman to even your CD-ROM drive on your computer, if it has music-playing capabilities.

Whichever model of headphone you choose, put it through the paces and get the one that sounds best to you.

MICROPHONES

Microphones are the electronic starting point of almost every recording you've heard. For this reason, it is important to purchase quality equipment and know how to use it.

MICROPHONE MECHANICS

As we have said previously, a microphone is a transducer. It transforms the sound pressure waves into electrical energy. The level of electricity, however, is very low and needs to be preamplified in order to be sent to other equipment. The microphone hookup on a receiver or tape deck usually contains this amplifier. From there, the signal is either recorded or sent to the speakers.

TWO TYPES OF MICS

Dynamic microphones are usually cheap and well constructed. They will reproduce sounds at a relatively good fidelity. A *condenser* microphone has a built-in preamplifier, and possibly a power source such as a battery. This makes for a better-sounding mic because the signal is amplified as close to the transducer as possible — eliminating any noise that may enter through a cord.

DIRECTIONALITY OF A MICROPHONE

We can break down mics into two more categories. *Omnidirectional* means that the mic will pick up sounds from ALL directions. This is good for reproducing the acoustical qualities of a room, but may also add too much background noise to a recording. The *unidirectional* microphone takes sounds from one direction only. These can be used for vocals, or backed up slightly to pick up an instrument.

CHOOSING A MICROPHONE

Buying a microphone seems like a no-brainer. Go to Radio Shack or WalMart and whip out the Visa for one of the three models they have. One is cheap, the other a few bucks more... But is that really all there is to it? Far from it.

First, think about how you are going to use the microphone. Will you be making professional recordings, or simply using it to take verbal memoirs? Will you need a very directional microphone, an omnidirectional mic, or even multiple microphones? Once you have a purpose laid out, then begin the research. Check out reviews in consumer, audio or studio magazines. Once you have a mic in mind, find a store that allows you to make a deposit to take it home for testing. Put the mic through its paces. Make sure there is no external noise creeping in from the cords or amplifier. If it passes the test, then buy it.

WRAP UP

Microphones are the beginning of the audio chain. Loudspeakers and headphones are the end. It is important that they are all high quality and do not introduce unnecessary noise or distortion into a system.

The technology of speakers is not as important as setting them up correctly. You bought your stereo in order to listen to the stereo effect of your favorite music. Make sure the speakers are actually set up to give you that effect. Otherwise, we might as well be listening to mono systems.

See Chapter 11 for diagrams and hints on how to hook up your loudspeakers, and Chapter 10 for choosing the sweetest-sounding speakers for your system.

CHAPTER 9
COMPUTER SOUND

"Someone yanked the sound card out of my mind, and now I can't hear myself think." John Adams

Guns are blazing and buildings are exploding as you type away, and suddenly your computer speakers spit out a friendly reminder: "Congratulations, Skippy! You have E-mail!" Computer sound has become an all-but-transparent sound technology; people who use computers on a regular basis tend to take computer sound for granted. But what makes those dwarfish speakers pound out such great sound?

With the right software, CD-ROM drives can be used to play your audio CDs, sometimes with better sound quality than your hi-fi! Sound effects can be added to your computer setup, designed to alert you to upcoming appointments and incoming E-mail, or simply to entertain you while you're working. Some computers can even be configured to speak, reciting information when you need to step away from the screen to do something else.

Let's examine how the sound card and speakers of a computer conjure up new acoustic worlds for our ears.

Sound has come a long way for computers, from the simple beeps of the built-in speaker to the current 64-voice, digital-wave synthesized MIDI-capable cards, 3D surround sound, and CD-quality digital sound effects (not to mention full CD-Audio to boot!). Whew! What an earful.

SOUND CARDS

Computers are digital. They are capable of spitting out ones and zeros, not analog waves. So, in order for a computer to send sound to its speakers, it has to use a digital-to-analog converter (DAC). These con-

NOTE: See Chapter 7 for more information on digital technology.

verters are very similar to the ones a CD player uses, except they reside on a computer card that plugs into your motherboard, or are located on the motherboard itself.

Now here comes the tricky part: How do you use digital sound? By converting the analog signal into a digital one, using a sampling rate, of course! In essence, the sampling rate gives you the time interval for sampling the sound wave amplitude.

16-BIT SOUND

Most sound cards on the market use a 16-bit DAC. This allows the computer to send a 16-bit binary value to the DAC every split second. It converts this to one of 65,536 voltage values that are used to drive a speaker's coil. In the next split second, another value is sent to the speakers. This recreates a sound wave the speakers understand. This is the standard for today's sound cards.

MIDI

The next thing to examine is MIDI. This stands for *musical instrument digital interface*. Many years ago, when instruments and recordings began to go digital, there was a need to make a piece of music that was recorded on one system sound the same on another. For this reason, a number of companies got together to form a set of common instrument sounds and the commands to drive them. This is referred to as *General MIDI*, or *GM*.

By taking the predesignated instruments, you can route them through any digital instrument (MIDI compatible), or even to your sound card. Sound cards at first used simple frequency modulation (FM) sound effects to replicate the instruments, but it wasn't very realistic. The current trend is using digitized instrument sounds. The sound card (or compatible software) will mix the sounds for you and play them through the speakers. On the Sound Blaster 32/64 AWE, this involves using the Advanced WavEffect (AWE) chip set. The difference in sound between FM and AWE is remarkable.

3D SOUND

This newer sound technology attempts to recreate spatial information using two or more speakers. Sounds are modified to make it seem as though an object is moving around you, with its tone and frequency changing as it moves.

If a car in a video game is rushing from left to right, or a jet is taking off away from you, the computer can recreate an acoustic approximation of what it would really sound like. This effect can be simple or difficult, depending on the situation. In real life, a sound fades into the distance when moving away and increases in volume as it comes toward you. There is also a shift in the *frequency* of sound.

As mentioned, you can use two or more speakers for 3D sound. With two, you can approximate the sounds on either side. But by adding a surround sound system, the experience comes closer to reality, with sounds coming from all directions.

WHAT CARD TO BUY

When it comes time to pick your sound card, what do you buy? Well, in this new age of electronics, what is hot today may be obsolete tomorrow. Planning ahead may help, though.

The Doppler Effect

The Doppler effect occurs when the frequency of a sound changes depending on its movement. If the object creating the sound is coming toward you, or if you are moving toward the object, the frequency (pitch) increases. Moving away, the pitch decreases. Imagine you are at an Indy car race in the stands. A car is racing around the track and the motor is at a constant speed. The frequency of sound from the motor is therefore constant as well. Now as it moves toward you, the frequency of the motor seems to increase. It will become an almost screeching sound. After it passes, the frequency beings to decrease. This effect is created because the object's sound wave is rushing toward you at an increased pace: i.e., the speed of the wave plus the speed of the car. Computer 3D sound can recreate the Doppler effect.

NOTE: SB32 requires a special add-on board to achieve MIDI capabilities.

What has to be the undisputed champion of sound cards is the Sound Blaster series (SB for short). Most other sound cards, usually called *SB-compatible*, are merely copies of the SB, or use similar chip sets. There's the Sound Blaster, Sound Blaster Pro, Sound Blaster 16, and Sound Blaster 32/64 AWE. The last two have digital MIDI capabilities and are the recommended buys.

The SB 16, 32 and 64 are about the only cards you can get nowadays. The others are obsolete. If you have the money, go for the SB 64 (just under $200 at most retail stores). The SB 32 goes for about $90. The SB16 is around $30. Some knock-off brands are as little as $10. In fact, some motherboards now have built-in SB 16 capabilities, negating the need to buy an add-on sound card.

The number after the "SB" name means different things. The Sound Blaster 16 can handle 16-bit digital sound. 32 means it can handle 32 different voices (for FM/MIDI playback). Finally, the 64 can handle 64 voices for MIDI. Note that all sounds cards above SB 16 are 16-bit with a special chip attached to handle MIDI and voices.

I recommend the SB64 AWE. It's cheap (and getting cheaper every month), has sound expansion capabilities, and comes with an optional CD-ROM/SB/speaker combo. If you want the cheapest computer sound possible, try a SB16 compatible card. Beware! Installing this piece of hardware can be the most frustrating computer upgrade you will ever make, due to potential conflicts within the computer. You may save $50 on the sound card only to end up giving it to someone else to set up.

Please note that these sound cards are specifically designed for IBM-compatible or Windows machines. Apple Macintosh computers come with impressive sound capabilities already built into the machines, so purchasing a separate sound card isn't necessary. Instead, focus on purchasing speakers and sound programs that will allow you to do what you want.

SPEAKERS

Good sound systems come with even better speakers. If you invest enough money in your sound card, you should buy good speakers to match. First there is *mono*, which is hardly used in computer sound. If you want minimal sound, get at least a pair of unamplified stereo speakers, which should provide about 4 watts of output power. They don't sound like much, but when these speakers are 2 feet from your head, they are fine for basic sound. Medium-end would be a pair of quality amplified speakers. Something from 20 watts to 60 watts would suffice. Upper-end would be a sub/sat combo. The highest level, which is the best setup, is surround sound. This has five speakers: three in front (left, center and right) and two to the sides. Some computer speaker manufacturers include a subwoofer somewhere for bass. The subwoofer can draw as much as 240 watts, and the side and surround speakers can hit 170 watts. This is more than most home stereo systems use. Another speaker option would be to connect the sound card's output to a home stereo system. Whichever speakers you choose, make sure they have magnetic shielding to prevent damage to your monitor.

NOTE: Just because you have a surround sound system does not mean your software will work with it. Make sure your sound system is compatible with the programs you are trying to use.

WRAP UP

Computers are becoming integrated into our home entertainment systems. It is best to learn *now* how the computer's audio can be made into an extension of your current stereo system. Besides, who wants to listen to a video game soundtrack on puny, unamplified computer speakers? Hook your computer up to your big tower speakers/amp and BLAST AWAY!

CHAPTER 10
BRANDS & CHOICES

"The QUALITY of sound, NOT QUANTITY of gadgets, should help you decide which model to buy" Unknown

So many choices, so little money. That's the technology bug at work. When was the last time you DIDN'T feel guilty about buying what you wanted, no matter the price tag? Making a well-thought-out, researched decision may help ease that guilt. That's what this chapter is for. At the very least, you will be able to look your spouse in the eye and say, "I think this subwoofer is the BEST deal. We should get it. Besides, it's on sale!"

QUESTIONS TO ASK YOURSELF BEFORE LOOKING

When considering a stereo system, look deep inside yourself and ask these questions. Make sure you give an honest answer and not an unrealistic one (if this is humanly possible):

- How do you intend to use this audio system? For music? TV sound? Recording?
- Where will the system be placed? Do you have enough room for it? Will the room need to be soundproofed if the volume control accidentally goes to max?
- What kind of music will you play? Will it be bass-heavy, treble-high or balanced? Does it require special equipment to reproduce? Will it reproduce the acoustics of, say, a concert hall?
- What level will you play? Will the higher volumes blow your speakers or amplifier? Should you get a higher-rated amp/speaker combo?
- Do you want "OLD" sound that's classical and has character, or new digital sound that's clean and charismatic?
- Do you want to hook it up to your home theater? Can you purchase certain components now and hold off on the extras until a windfall?

When you have the answers to these questions, it is time to start laying out a system that matches what you want.

STEPS TO DECIDING BRANDS AND MODELS

You are taking the first step right now. You are learning the nomenclature of audio. Learn the jargon and you will be able to communicate your needs to a salesperson.

RESEARCH

Research your audio area of purchase. If you want a surround sound system, look up all the information you can in fliers, books, ads and Websites. If you still have questions, E-mail, call or write the manufacturers and give them your query (preferably in that order). See if one brand or model gets consistently high ratings. Does it match your criteria for a system?

Read Ads: Ads are a valuable commodity for helping you learn about electronic products. Glean as much technical information from ads as possible. While you are at it, try to get a general idea of the item's cost. Most stores list a manufacturer's suggested retail price. You can pretty well ignore it because in today's competitive market, the real price is usually 10-50% lower than advertised.

Compare Features: Take your time to find out what each feature really means and does. Do you really need to pay 30% more for more buttons on the front of the receiver?

Ask Friends: Ask your friends about the brands they've bought and the experience they've had. A friend would be the one to ask if it's worth spending an extra $200 for a higher-quality set of speakers.

Find a Knowledgeable Salesperson: They ARE out there. Look for salespeople who do not work on commission, and their brains. Test them! Then see if they can help you understand the features you don't understand.

Get More Product Information: You can do this by visiting the manufacturers' Internet Websites, or by calling their 800/888 numbers. See the Appendix for a list of Websites, phone numbers and addresses. Some stores will also supply color brochures.

BRANDS

The most common questions people ask regarding electronic equipment are, "Which brand should I buy?" and "Which model should I buy?" This is like asking someone else, "What color do I like?" It is purely a subjective choice. There is no simple answer. So I will tell you the questions to ask to help you make your decision:

HOW DOES IT SOUND?

If a system doesn't grab your ears, shoot your emotions up and down like a roller coaster, exciting you down to your tippy toes when first listening to it in a store, it certainly will not catch you in a less-than-acoustically-perfect environment, such as your home. Audio stores have sound rooms that are specifically designed to sound super. So make sure the music coming from the system you want to purchase moves your very soul before purchasing. For criteria on how to judge the sound, see *Rating How a System Sounds* in this chapter.

OTHER QUESTIONS

- How does the brand/model rate in the audio and consumer magazines?
- What are others saying about it?
- What is the highest model in that range you can realistically afford?
- Will you be satisfied with this purchase or always wondering if you should have went with another brand?
- Does the manufacturer or retail store offer a good warranty and can deliver on that promise?
- How does it look? This is important, but make sure it is the last question on your list.

HINT: Salespeople will often turn up the volume slightly on the model they want you to buy. Higher volume always sounds better. The model maybe worse-sounding, but that little bump in the volume makes it sound superior to other models. Tricky!

BRANDS TO LOOK FOR

There are many high-quality manufacturers that produce stupendous products in the audio market. Here are a few of the better choices:

Advent, Aiwa, B&W, Bose, Cambridge SoundWorks, Denon, Harman-Kardon, Infinity, JBL Consumer, JBL Pro, Linn, Marantz, NAD, Nakamichi, NHT, Onkyo, Paradigm, Polk, Rotel, JVC, Sony (especially the Elite! Series), Technics and Yamaha.

Try to steer away from consumer reports-type brands for high-end audio. These components are usually low-quality equipment with a fancy paint job and useless trinkets. The manufacturers are more concerned about consumerism than audio. You know which brands I am referring to. These are the ones that are always on sale at you local K-Mart or WalMart.

RATING A BRAND

When push comes to shove, brands usually don't matter; it's the model differences that count. However, when going shopping for new equipment, you may want to rate the brands and models yourself. Try to rate these criteria:

Receivers: FM reception, ease of use and power.

Loudspeakers: Bass/treble response, minimum power.

Tape Decks: Features, player quality, recording quality.

CD Player: Cost, service record.

RATING HOW A SYSTEM SOUNDS

Always listen to a system before you purchase it. Independent ratings are great, but people's tastes are different, especially when it comes to sound. A rating may direct you, but don't use it as a sole decisive factor.

The most important maxim to know when rating a system is that during the first few moments you listen, you are most sensitive to little quirks in the sound. This is because your senses are most sensitive when first exposed to a stimuli, and tend to adapt quickly. So, if a system doesn't sound good when you first hear it, it is not going to sound any better three months later.

The next datum to remember is to always use a recording that you are familiar with in order to compare systems. Bring your favorite CD with you and ask to play it on various systems and speakers.

Now listen for sections of the audio spectrum that sound "hot" or overly loud. Is there is a particular section that will allow you to test the subwoofer? Go to it, and see if the booming is overpowering. Set a midrange passage and make sure the voices sound crystal clear. Do the tweeter reproductions hurt your ears, or are they pleasantly filling in the sound for you?

Make sure the unit is reproducing a stereo image as well as possible. Close your eyes. Can you picture the sound stage before you? Are the drums on the left, the singer centered, and a guitar to the right (if this is how the stage was recorded)? If not, are the speakers positioned right? (see Chapter 8 if not). Does each instrument sound like the real thing, or does it sound tinny or muffled? Does the unit reproduce musical notes well? Can you hear each one as a separate entity? In other words, is the instrument mixing so bad that you cannot tell a drum from a bass guitar?

Now that you have a pretty good feel for the systems and have narrowed your choice somewhat, begin stringent tests with music that is familiar to you. Blast the volume and see if there is any distortion under heavy sections. Do the speakers clip out or distort when every instrument in the recording is blaring? If so, the system is underpowered. Is there distortion under low to mid loads? This is also a symptom of a low power amp or mismatched speakers. Look for any combination that creates distortion. Play around and see which system can hold up to your tinkering.

Chapter 10: Brands & Choices

Quite a procedure just to choose a piece of audio equipment, eh? But then again, it is your hard-earned cash at stake.

WHERE TO PURCHASE

There are basically two types of stores in which to buy audio equipment. Electronics boutiques (such as Circuit City and Best Buy) and department stores are fine for lower-quality combination or rack systems. A/V specialty stores have much more knowledgeable staff and tend to carry better brand names. Whichever type of store you choose, make sure it meets the following criteria:

- They stock and display the item you want.
- They service what they sell.
- Your friends have recommended them.
- The salespeople don't work on commission. This is difficult to find, however.
- They employ salespeople who actually answer your questions and are courteous and prompt in helping you.
- They are consistently busy. Watch the store for a few months and see if people actually return. It's usually a sign of good prices and good service.

Try to avoid any of these types of stores:

- A fly-by-night operation tucked away in the corner of a strip mall.
- Stores that jack up prices then put them "on sale." This is a sure sign that you are about to walk into the lions' den during feeding time.
- Mass-marketers such as K-Mart, WalMart, Target, etc., who offer brands such as GE and Sanyo. These are usually considered economy products. An example would be those shelf stereos that supposedly pump out great sound for their size. If you are extremely tight on money, then by all means purchase them. If not, consider looking elsewhere as these stores will rarely offer any sales or service help for the products they sell.

The greatest advice I can give is to find a shop where you feel comfortable and are able to actually talk to a salesperson who understands you. It shouldn't matter if the store is an electronics boutique or audio specialty store. What matters is the service you get.

PURCHASING STRATEGIES

After reading this book, you should generally know what you want to buy. You have X amount of dollars to spend on what you want. Now it's time to balance your audio budget with your audio needs and wants. See Figure 10-1.

Once all of the electronics jargon is under your belt, and you are pretty sure about what each feature truly does, start perusing local ads. Cross out the stores that meet the "avoid" list. Go to the few lucky stores left on the list and see how they feel; dishonesty has a nasty odor, and I am sure you will be able to smell it.

Make a note of how fast the salespeople try to help you. If it takes more than a few minutes, don't stand around: leave. When someone does approach you, start asking questions. If you don't get your questions

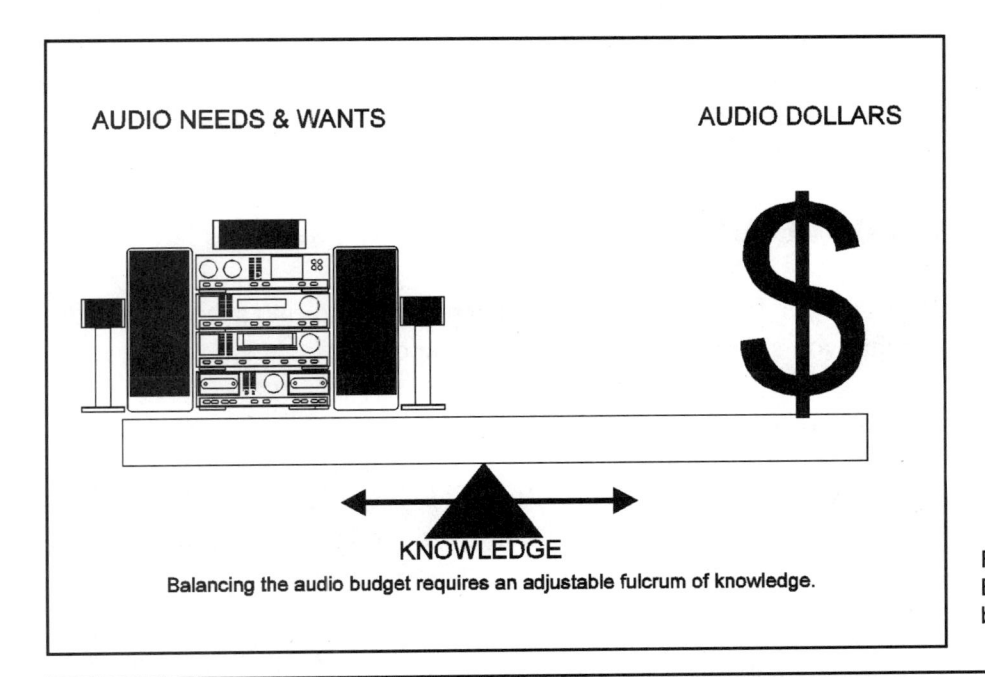

NOTE: Electronics boutiques rarely have their sound rooms set up correctly. If you are into heavy comparison, try an audio specialty store; they are more likely to have the equipment set up for optimum sound.

Figure 10-1. Balanced budget.

Balancing the audio budget requires an adjustable fulcrum of knowledge.

Chapter 10: Brands & Choices   117

answered, cast your buyer's vote and walk out. No use in wasting your time or, more importantly, your money.

Before walking into a store, you should know which brand you want and what to expect to pay for it. Make sure you walk out with that goal intact: take the attitude that you are there to BUY something, not to be SOLD something. The right salesperson for you will sense this and be more than willing to help. After all, it's easy money, right?

OTHER PURCHASING TIPS

- Always walk into a store knowing the brand and model YOU want to purchase. If the store doesn't have the item, try somewhere else. Stores are sure to try to sell you an overstocked or nearly obsolete item otherwise.
- Look for a price guarantee. Most electronic boutiques will pay you the difference if an item becomes cheaper within a set period of time. Example: You purchase an A/V receiver for $800 in August. Then the store runs a $700 special in September for the same receiver. The price guarantee allows you to go to the store with the receipt and get a $100 refund.
- Find an establishment that covers the product's warranty on the premises.
- Beware of service contracts. An alternative is to have a *slush fund* put aside for all your appliance repairs.
- Become a purchaser, not a consumer.

RECOMMENDATIONS

Now that you know how to research a product and find a store, let's turn to what you should look for in each component.

COMBINATION, SEPARATE OR RACK?

If you are tight on space and don't mind lower-quality sound, try a combination component system. Look for a model with simple controls, detachable speakers and a remote.

If you have read this entire book, you should have enough knowledge to know how to pick out a separate component system. Customizing your very own surround sound package can be very rewarding. Make sure to allow for a few extra bucks in the budget, though.

If you are still not comfortable with your knowledge of audio equipment, but want something a bit better than an all-in-one unit, try a rack system. You may get lucky and get a good package. At least you will have the choice to expand later. Make sure you get an A/V receiver, CD player and dual-tape deck in the deal.

SYSTEM STRATEGIES

Once you have decided which type of system to purchase, it is time to start mapping out exactly what you need. A combination or rack system is a no-brainer, so we will focus mostly on separate components:

Evaluate Your Current System: Take a serious look at the components you already have at home. Do you have an A/V receiver? Do your speakers met your needs? Is the amplifier large enough to power quintuplet speakers?

Try to see how each component relates to the rest of the system. Remember the saying, "A chain is only as strong as its weakest link." If a particular component is causing the audio chain to break down, then you should replace it.

Make a Wish List: Everyone has an ongoing, got-to-have-it equipment list. After you evaluate your system, you will want to add all the weak links to the list.

One Component at a Time: Figure out a budget and see what improvements you can make. Buy one item from your wish list per paycheck or per month, depending on your needs and resources. Build your new system one component at a time. Start with a better A/V receiver with Dolby Pro Logic, and preferably some Dolby Digital.

RECEIVERS

Your receiver is the center of your audio solar system. Each subsequent device revolves around it. So it is important for the receiver you choose to be able to handle each task that comes up. This includes anything from earth-shattering power to satellite speakers. An A/V receiver is the most universal choice, as it gives room for expansion and the ability to listen to stunning movie tracks as well as music. Here is what to look for:

- Try to get at least a Dolby Pro Logic receiver. If you can afford it, get the Dolby Digital model or at least a Dolby Digital-ready unit.
- Make sure it has various surround sound modes such, as Jazz Club or Concert Hall.
- Make sure the controls and setup are relatively simple.
- There should be three or four audio-only inputs, and two or more A/V jacks for VCRs, DVD, etc.
- If you are eventually going to hook up a DVD or DTV, DAT or MiniDisc, your receiver will need digital inputs. There are two types of these inputs; so if you, can make sure the receiver has at least one of both for compatibility.
- If your video equipment has S-video connections, make sure the receiver accepts these. Try to get a receiver with S-video connections for future use.
- Some receivers have onscreen control. Try to find one for convenience.

LOUDSPEAKERS

The most obvious thing to look for is a speaker set that sounds good to *you*. This would be speakers that make *your* music sound stunning, not gritty as they reach their input limits. Get the highest-quality speakers you can afford. Remember, this does not necessarily mean the most expensive speakers. Try to find loudspeakers that:

- Are compatible with your amplifier. If the amp manufacturer suggests an 8-ohm driver and a maximum of 150 watts, do not exceed that.

- Have quality cabinets.
- Have large-enough drivers to recreate the acoustic effects you want.
- Have magnetic shielding, especially if they will be placed near a picture tube.

CD PLAYERS

Most new CD players sound pretty much the same, so it doesn't really matter if you spend $150 or $500: the quality difference is negligible. It is more important to get the features you want and a CD changer that will carousel the disks. If the CD player will be moved around or bumped regularly, try to get a unit with bump resistance.

TAPE DECKS

When looking for a tape deck, make sure it:

- Contains Dolby B and C noise reduction.
- Has the ability to recognize various kinds of tapes.
- Is a dual deck.
- Has auto-reverse.
- Has low flutter, and high-accuracy recording and playback capabilities.
- Has headphone volume controls, if needed.
- Has at least three audio heads.

WRAP UP

You would think that while spending $3,000, people would make a well-thought-out decision. Actually, most people are hit-and-run consumers. After three minutes of listening, they plunk down a fortune. There is another type consumer, one who always takes another person's advice as the holy word, then runs out and buys without a listen. Don't be either of these types of audio buyers. Research a piece of equipment to death. Find a reputable store to purchase from, and make sure they service what they sell. Last of all, make use of the knowledge you've gained from this book.

HINT: If you want a sub/sat combo, make sure you purchase the subwoofer at the same time as the small drivers. They will not work as separate entities.

Caveat Emptor

Most electronics stores are reputable and honest. However, there are still sound snakes out there. Here are some of the tricks they use:

A store will advertise a home theater receiver for what seems like an unbelievably low price. It is likely that the item is about to be replaced by a newer, shiner model. Or the receiver may not be a surround sound receiver; don't always assume a home theater receiver has Dolby Pro Logic or any other surround sound system built in. In fact, I have seen receivers advertised that were not surround sound whatsoever, yet displaced the Dolby Pro Logic logo. On careful examination, the Dolby logo was part of another ad in proximity to the non-surround sound receiver.

Another trick: while you are looking at a pair of speakers the sales representative wants you to buy, they will switch on a subwoofer. This will naturally make the system sound better.

One problem I recently ran into is this: I purchased an item that was dead on arrival. After trying to return the item, I was informed there was a no-money-back policy, only credit toward something else. Now, why would I want to risk getting another dead item from the same store?

The last little illusion some really disreputable dealers use is to "bait and switch." This means that after your selection is made, they go back into a "warehouse" and get the item for you. Upon getting it home, you may learn that this is in fact a not the model you chose but instead a discontinued model. Buyer beware!

CHAPTER 11
HOOKUPS & ACCESSORIES

"It didn't have enough power, so I REWIRED IT!" Tim Allen

Audio equipment and accessories are similar to Lego blocks: they both contain the basic building blocks to help you assemble your toys. In audio, we are dealing with cables, connectors and, of course, the equipment itself. Here we will explore the purpose of these hookup devices and how to use them.

It is almost impossible to explain the hookup of every audio device on the market. Instead, we will teach you the basic principles of sorting out the spaghetti of wires at the back of your components. This includes common hookup techniques and parts used. Knowledge of how something works is always more important than only knowing how to hook up, say, a JVC receiver.

NOTE: The information in this chapter should not supersede the manufacturer's manuals; this chapter is merely a guideline.

WHAT YOU NEED

The first item you need is patience. Despite modern movements toward more user-friendly equipment, there still remain hookup situations that will send a sane man to the loony bin. Sit back and leave it alone for awhile if things aren't going right. No use in blowing out your thousand-dollar speakers because you lost your cool.

The other items you will need are listed here.

CABLES

Audio equipment uses many different types of cables. See Figure 11-1. Some send the signal from one section to another. Thicker cables pro-

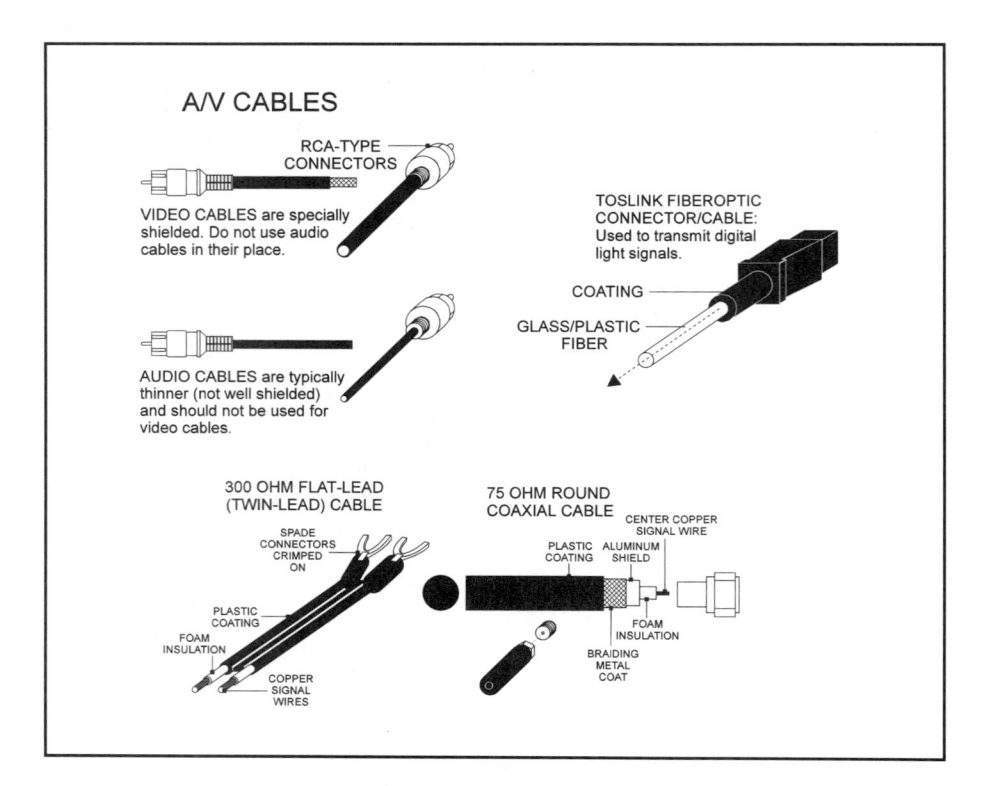

Figure 11-1. Common cables used in home audio.

vide power to the speakers. New digital cables transfer billions of sound bits.

Audio: Audio patch cables usually contain an RCA-type connector on both ends. They provide a path that allows the various sound signals to travel around your system. The quality, cost and materials can vary. They provide good protection against outside interference.

Video: These cables are highly shielded. They are not the same as audio cables, and should not be used interchangeably. RCA connectors are used on both ends.

Coaxial: This type of cable is used to carry an RF television signal. The newer digital audio signals are also carried over coax.

Fiber Optics: Thin strings of glass or plastic fibers are used to carry digital signals between components. They sends pulses of light down the line; which represent bits of information. Also called *Toslink*.

Speaker Wire: The type of wire used for speakers depends on the power load and the way the wire is being run. This can be anything from a thin, low-power pair of speaker wires to ribbon cable that can be hidden against or under baseboards. Some people even use 12-gauge Romex, which is usually used to run power to outlets in homes.

CONNECTORS

The shape, type and materials that make up the connector vary with its purpose. All connectors should provide as little resistance as possible, and be coded or marked somehow for easy hookup. See Figure 11-2.

RCA: These are the most commonly-used connectors in an audio system. They are usually on the ends of an audio or video patch cord. Some are gold-plated to provide a noncorrosive connection. If they are made of steel, make sure you clean them often to remove corrosion;

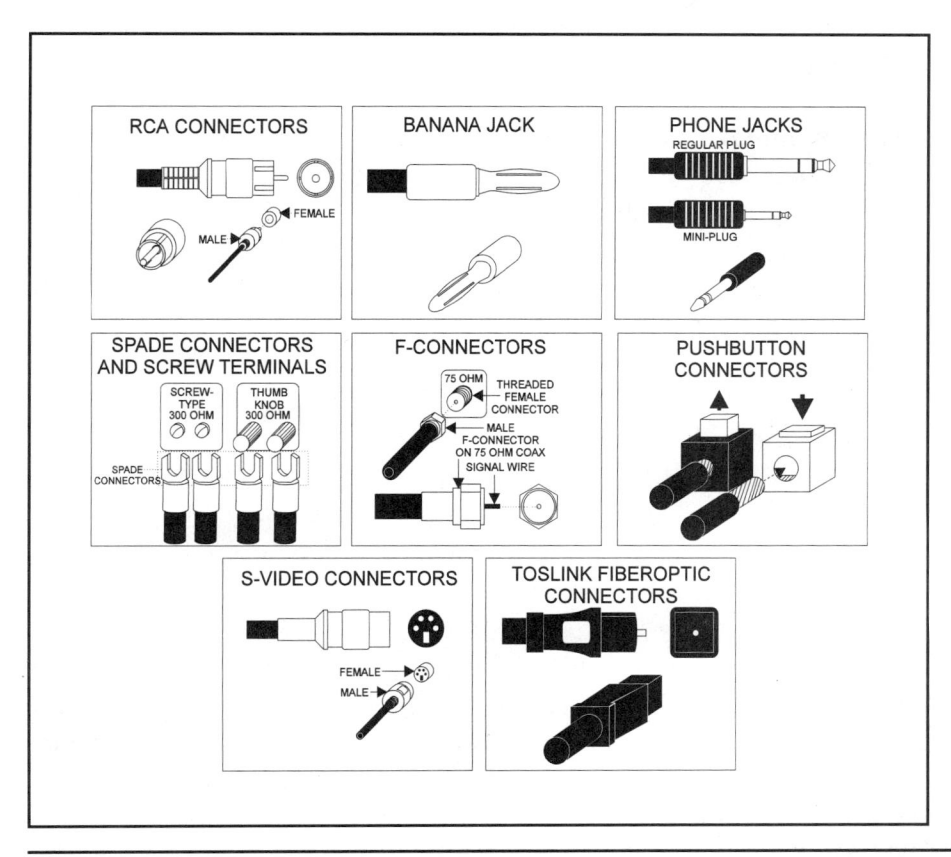

Figure 11-2. Common connector types used in audio systems.

Chapter 11: Hookups & Accessories 125

otherwise the sound quality will degrade because the residue on the connector causes excess resistance. RCA connectors are also used for digital signal patch cords.

F-Connectors: These are used in conjunction with coaxial cable. They attach RF cables to audio and video components.

Digital Connectors: New digital devices such as DVD players use one digital connection instead of six analog connections to transmit audio information to the receiver. There are two types of these. *Optical connectors* are on the end of the Toslink cable. They look like small square plugs. One connection goes into the DVD player and the other into the receiver. *Coaxial connectors* look like RCA connectors. Some machines have them to save on manufacturer design costs.

S-Video Connections: These connectors have five little holes, or five pins, on a black face. The cables carry the video signal, separated into component parts. It provides a superior way to route video around your entertainment center.

Spade, Banana or Push-button Connectors: Each receiver and speaker combination uses different types of connections. One is the *push-button* type, with a hole in the middle for the stripped speaker wire. Then there is a spade connector, attached to the receiver and speaker with a screw or thumbscrew. Some audio components use *banana plugs* for easier connections.

WHAT EACH CONNECTION ON YOUR RECEIVER IS FOR

The A/V receiver is the hub of your home entertainment center. Each component has several wires running into it. From it, signals are processed, amplified and sent out to your speakers. You control the whole show with the receiver's remote.

A typical modern receiver has about 50-75 connectors in the back. See Figure 11-3. Some people are driven crazy at the mere sight of these connectors. There are so many tasks for the receiver, with each device

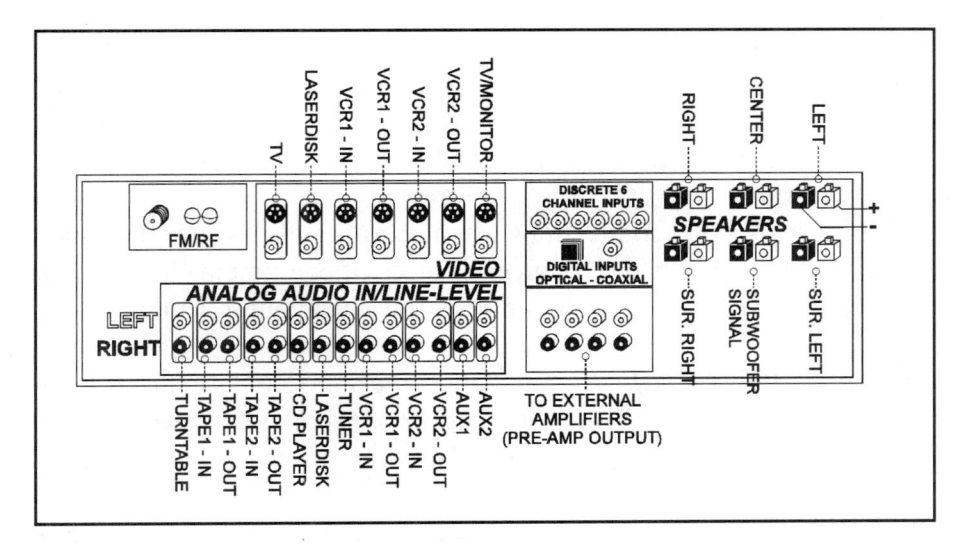

Figure 11-3. Connections on a typical A/V receiver.

wired to it. Unfortunately, you may find all the cables and connectors intimidating. Don't let them get to you.

Most receivers are laid out in a logical manner, color-coded and labeled. It is not the ultimate user-friendly, plug-and-play consumer product on the market; but with patience and a tiny bit of hookup knowledge, you'll be a pro audio installer before you know it.

TYPICAL CONNECTIONS ON AN A/V RECEIVER

Let's first divide the connections into sections and describe what each does, and which cables and connectors are used:

Analog Audio IN (Line-Level): This is the most commonly-used audio connection. An RCA-type connector is used to route analog audio signals to and from components and the receiver. They are usually used in left/right pairs. Some items, such as tape decks, require two sets; one for input and one for output. Some of the components that can be connected using these lines are CD players, tape decks, turntables, VCR audio, tuners, auxiliary devices, digital satellite audio and laserdisc audio.

Video IN/OUT (Composite Video and S-Video): Devices that send a video image to and from the A/V receiver can be connected using either

Chapter 11: Hookups & Accessories

a composite signal (RCA-type cord) or S-Video. The S-Video provides better separation of the video signal for better quality pictures. Components that use these connections are laserdisc video, television or monitor, VCR video and digital satellite video.

HINT: Use S-Video connections whenever possible. The signal is not separated and combined through each piece of equipment. This makes for a better final product once the signal reaches your TV/monitor.

Speakers OUT (Amplified): Since most receivers contain internal amplifiers for the speakers, you can simply connect the two leads to each speaker without running them through another piece of equipment.

Digital Audio IN: Newer audio devices use a single digital connection to reduce the amount of wires behind the receiver. This is either a special optical fiber connection called a Toslink, or an RCA connector with a coaxial cable attached. Devices such as DVD players, DAT, MiniDisc, digital satellite audio, laserdisc audio and (soon) digital television use this connection.

Line-Level OUT: This is also called *preamplified outputs*. These lines are used if you want to power your speakers with external amplifiers.

6-Channel Discrete Inputs: If you have a Dolby Digital-ready receiver and a DVD player, you will have to use six analog lines to provide an audio signal to the receiver. If you have a Dolby Digital receiver, simply use one digital coax or one Toslink cable.

FM/RF: These inputs are used to wire an antenna signal to the receiver. They may use an F-connector and coax, or a twin-flat cable and two screws for hookup.

HINT: It is always better to use a digital audio connection between equipment. This reduces unnecessary system noise.

HOOKING UP COMPONENTS

Now that you understand the basic building blocks of audio, it's time to build! We will look at how various components are connected to the A/V receiver (the heart of the stereo system). After looking at Figure 11-4, you are probably thinking, "NO WAY CAN I POSSIBLY HOOK UP

GOING FOR THE GOLD

Good connections are important to any electronics device; especially one that carries high-frequency signals such as those found in audio and video components. A tight, corrosion-free junction between the cable and the equipment is the key.

Some cable manufacturers use gold plating on the surfaces of the connectors. High-end equipment manufacturers are also plating the connections on the back of equipment. Why are they using gold? Gold is highly resistant to corrosive elements. It stops a buildup of crud that can cause an imperfect connection and lousy sound. The problem is, the connectors are expensive!

Do gold connectors make your system sound better? No. They stop audio signal-killing corrosion from interfering, that's all. So, you have a choice: clean the regular connectors often, or spend a fortune going for the gold.

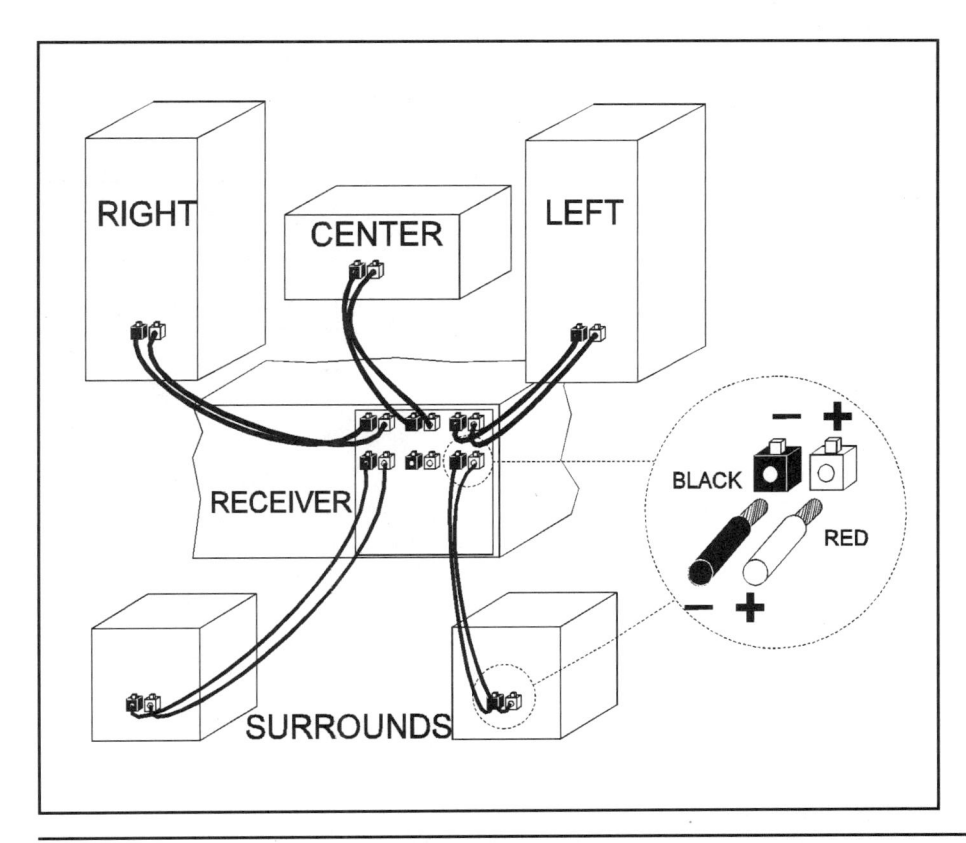

Figure 11-4. Hooking up speakers.

Chapter 11: Hookups & Accessories

THIS MONSTER!" When we break it down into each component, you will see just how simple it is. So take a deep breath and dive in.

HOOKING UP YOUR SPEAKERS

Figure 11-4 shows a typical speaker hookup diagram. The most important item to remember while connecting speakers is to obey the polarity of the speaker and the receiver or amplifier. Hook the positive leads to the positive connectors, and the negative leads to the negative connectors. This is very important because if they are reversed, you will not get a stereo effect; the speakers will be working against each other.

Try to make the leads to the speakers as short as possible to prevent unnecessary noise in the system. If you want to hide the wires, use flat speaker wire and run it under the baseboards or carpet edge. Push-button connections are pictured, but you can easily use banana plugs or a spade/screw combination. Banana plugs will wear with time and provide less contact; however, they are handy if you need to connect and reconnect the speakers often. If you want a death grip connection, use high-quality spades and screws. Push buttons are in-between.

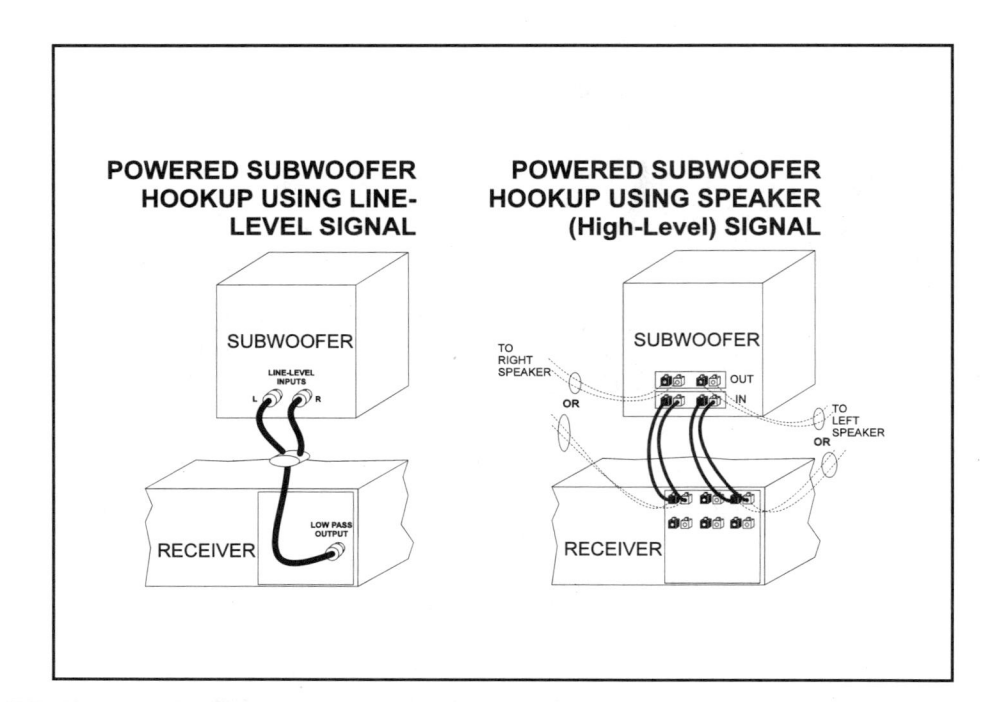

Figure 11-5. Hooking up powered subwoofers.

HOOKING UP A SUBWOOFER

If you have a subwoofer with a built-in crossover and amplifier, then you simply have to connect the low-pass output on the receiver to the subwoofer's inputs. See Figure 11-5. If the subwoofer has no amplification, hook up the amplified left and right channels (or spare channel) to the left and right high-level inputs on the subwoofer. You can then connect the left and right speakers to the subwoofer's high-level outputs, or run them from the same wires on the receiver if it is a shorter path.

HOOKING UP A CD PLAYER TO YOUR RECEIVER

The receiver will have a left and right jack for input from the CD player. See Figure 11-6. Connect an audio patch cord set with RCA connectors to the CD player's line-level outputs. Easy!

HOOKING UP A TAPE DECK TO YOUR RECEIVER

Because a tape deck records and plays from the same device, you will need two left/right pairs of audio patch cables with RCA connectors.

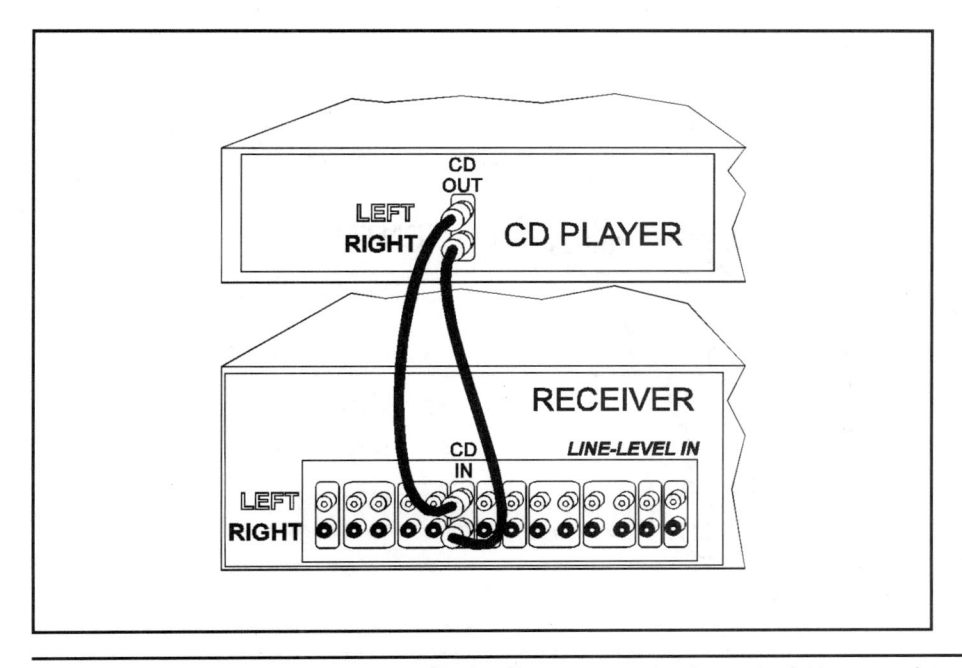

Figure 11-6. Hooking up a compact disc player.

Chapter 11: Hookups & Accessories

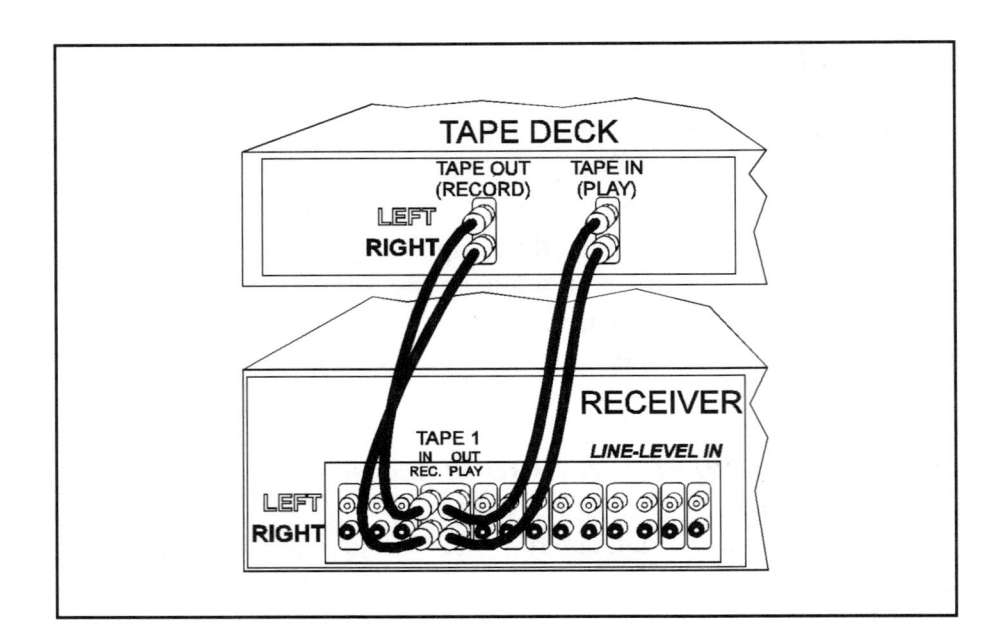

Figure 11-7.
Hooking up a
tape deck.

See Figure 11-7. Run the first set from the tape deck's *Tape Out* (record) to the receiver's *Tape In* (record). Now run another set of left/right cables from the tape deck's *Tape In* (play) to the receiver's *Tape Out* (play).

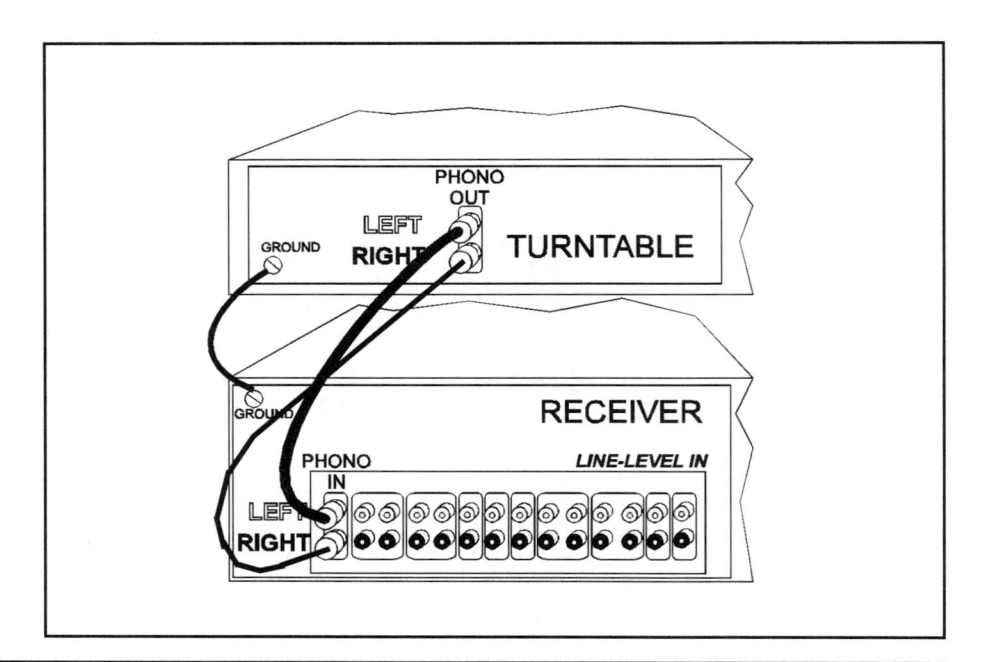

Figure 11-8.
Hooking up a
turntable.

132 Howard W. Sams & Company **Complete Guide to Audio**

HOOKING UP A TURNTABLE TO YOUR RECEIVER

Connect the phono line-level *out* (on the turntable) to the receiver's phono line-level *in*. See Figure 11-8. A ground needs to be placed between the turntable and receiver to prevent noise. You can use a spade connector and a piece of speaker wire.

HOOKING UP A VCR TO YOUR RECEIVER

Some people use RF connections to run combined video/audio signals between equipment components. The problem is that each piece of equipment splits and recombines the signals, causing degradation. It is better to use a path for the video and a path for the audio.

Figure 11-9 shows the proper way to connect a VCR using these separated paths. Hook up the audio as you would a tape deck, and connect either an S-Video cable or a composite video cord between the *VCR video* and *receiver video*. Try to use an S-Video line whenever pos-

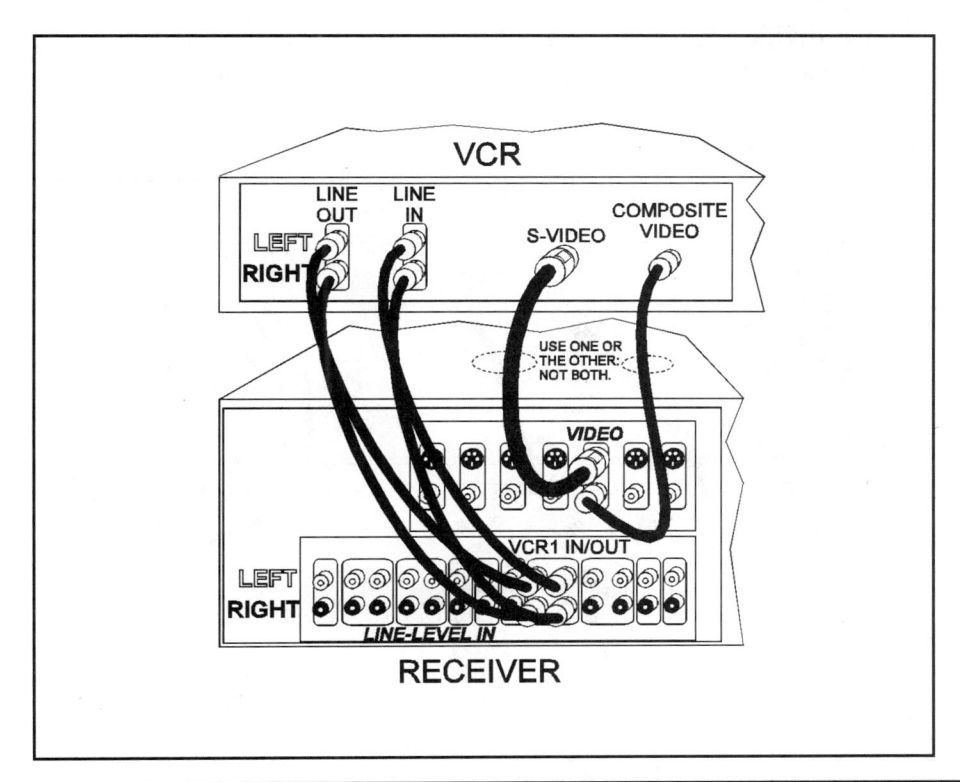

Figure 11-9. Hooking up a VCR to your receiver.

Chapter 11: Hookups & Accessories

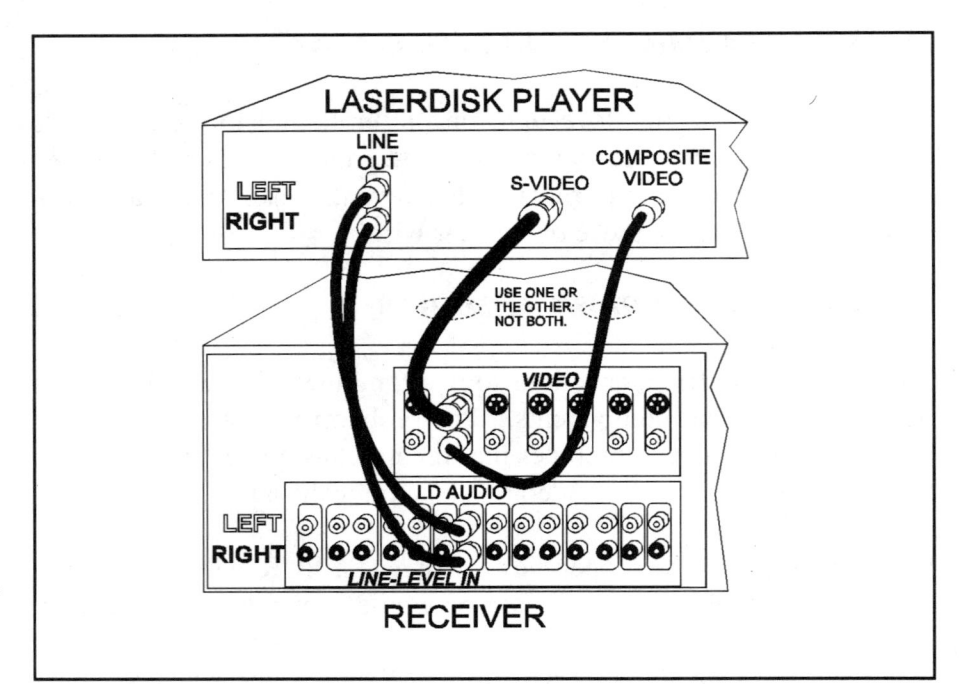

Figure 11-10. Hooking up a laserdisc player to your receiver.

sible; it divides the video into its separate components and prevents signal loss from the combining and recombining of the video signal components.

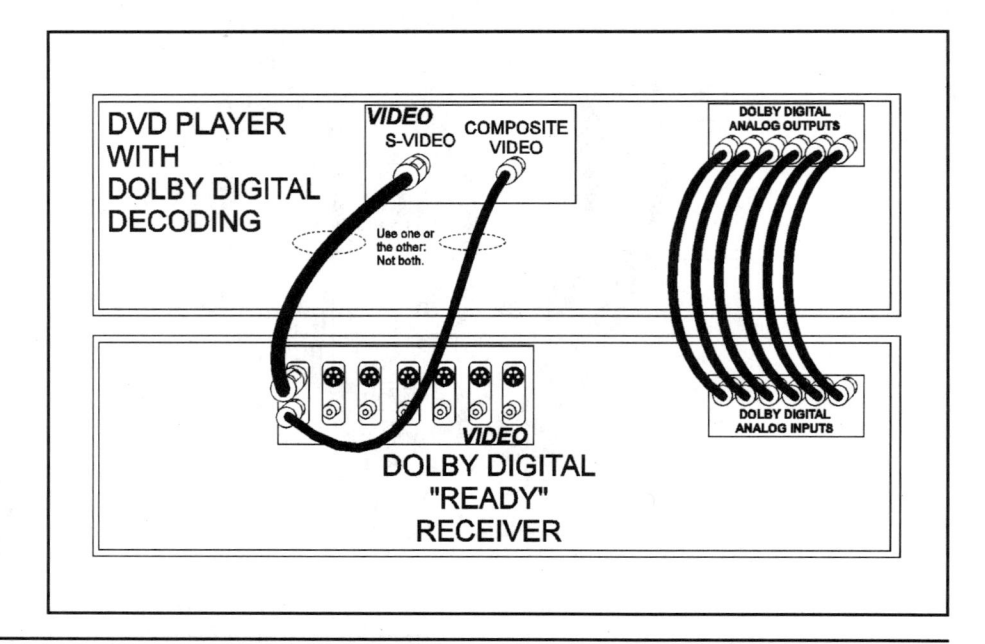

Figure 11-11. Hooking up a DVD player to a DD-ready receiver.

HOOKING UP A LASERDISC PLAYER TO YOUR RECEIVER

Laserdisc players are slowly being replaced by DVD. But if you have a player, here is the hookup method. See Figure 11-10. As with the VCR, try to use S-Video.

HOOKING UP A DVD PLAYER TO YOUR RECEIVER

DVD players provide the receiver with a Dolby Digital audio signal. The problem is that older receivers don't have the Dolby Digital decoding circuitry built in. However, an older receiver may be a Dolby Digital-ready receiver. In this case, follow Figure 11-11. This uses the six discrete inputs/outputs to run an analog audio signal to each of the six channels in the receiver.

If your receiver has a built-in Dolby Digital decoder, then you can hook up the player with a digital audio connection. See Figure 11-12. You can use a digital coaxial cable with RCA-type connectors, or the higher-cost Toslink fiber optical connection.

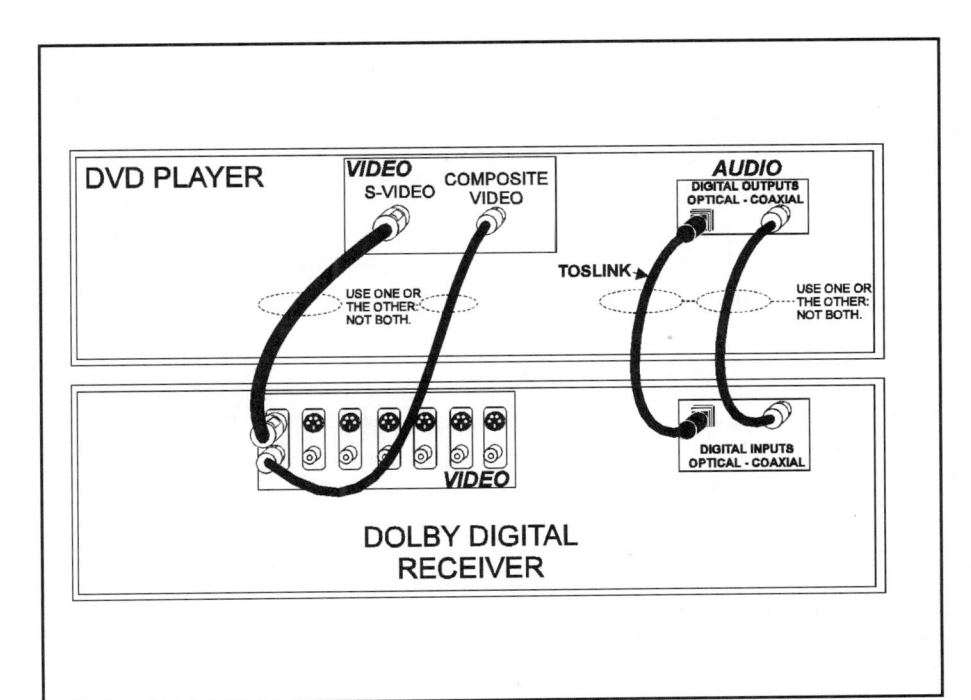

Figure 11-12. Hooking up a DVD player to a Dolby Digital receiver.

Chapter 11: Hookups & Accessories 135

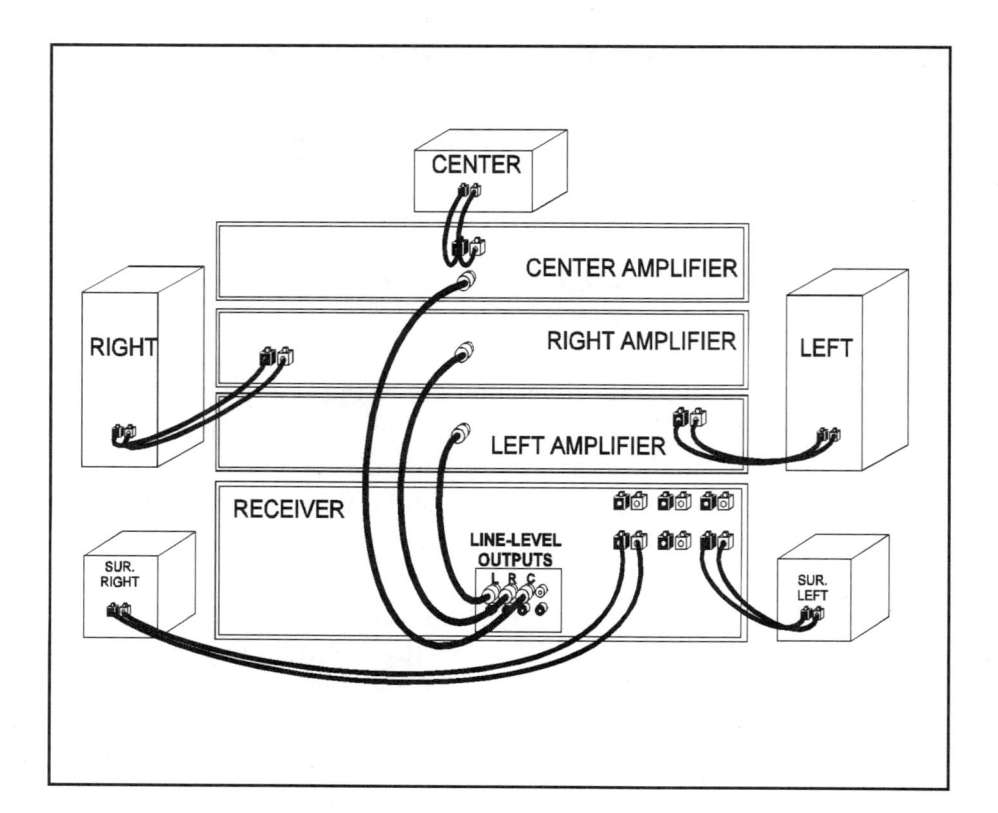

Figure 11-13. Hooking up external amplifiers to a receiver.

HOOKING UP EXTERNAL AMPLIFIERS

If your receiver just can't deliver the power you want, you can connect it to external amplifier. See Figure 11-13. By using the receiver's preamp outputs, you can theoretically connect an external amplifier for each channel. Start with the left, right and center channels. The surround channels typically use less power than the fronts, so you can add these later.

HINT: Companies such as Monster Cable Products, Inc., sell color-coded and combination cables for trouble-free installations.

WRAP UP

Don't ever get discouraged when installing your audio equipment. There are tons of wiring to deal with, which can easily get mixed up if you don't pay attention to what you're doing. If you are getting tired of fumbling with all the filaments, sit back, get a piece of paper and draw out each connection. Mark cables with masking tape if you have to. Take a look at the diagrams in this book again and look for the logic behind the audio assembly (if there really is any). Pretty soon, you will be enjoying those hard-earned tunes and movie soundtracks.

CHAPTER 12
FEATURES

"... it had hundreds of knobs, of which two actually worked: the on/off and the station selector." Bill Cosby

Now that you have your sleek new A/V equipment hooked up and ready to fly, it is time to figure out what each little doodad and gizmo does. This chapter gives you a brief rundown on features of common equipment.

If you are trying to decide on a purchase, don't always use the features list of the unit as an end-all to your choice. Remember, the sound is more important than the dials and buttons. If you want a more detailed list of features for a particular piece of equipment, do some research on the Internet or order more information from the manufacturers. Beware: most companies are quick to LIST the features, but won't tell you what each does. They merely make it seem like the component is a better value because of all the features it has. In fact, most features are simply something to keep you busy; but you are wise to that now. Go for the quality of the component, not the quantity of gadgets.

RECEIVERS

Dolby 3 Stereo Mode: If you only have two front speakers, you can use this to cancel the surround sound.

Dolby Pro Logic Circuitry: The circuitry needed to decode a Dolby Pro Logic signal.

Dolby Digital Decoder, AC-3: The DD decoder is built into the receiver.

DSP with Acoustic Environments (Modes): This will simulate the acoustics of a jazz hall, concert hall, stadium, you name it.

Front A/V Inputs: If you want to occasionally connect a video game console or camcorder to the receiver, these front jacks are handy.

Input Selector: Lets you choose between the various signal sources hooked to the receiver.

LED Display: This will display everything going on, including levels, sources, radio stations, etc. Most new receivers use only one LED display for everything.

Onscreen Display: Most high-end A/V receivers send a video signal to the television to provide an onscreen menu and status system. It's great if you can't see the receiver controls from the comfort of your couch.

Outlets: Receivers typically include plug-in outlets for devices such as external amplifiers, etc. This allows all of the equipment come on at once.

Radio Station Presets: To call up stations without having to go through each number manually, you can use programmable buttons called *presets*.

LOUDSPEAKERS

Active/Passive Subwoofer: An active subwoofer has the amplifier and crossover built in. The passive requires an external amplification source.

Bi-Amp or Bi-Wiring: Some speakers allow you to hook separate amplifiers to each driver. In this case, there are more than one set of wires to hook to the back of the cabinet.

Fuse or Circuit Breaker: Because of the heavy current going through some speakers, a fuse or circuit breaker is installed on the back to prevent damage to the speaker coils.

Magnetic Shielding: Speakers placed near a picture tube require some kind of magnetic shielding.

CD PLAYERS/CHANGERS/JUKEBOXES

Bump Resistance: If a CD player gets jarred, it will move the laser, causing a skip in the music. Circuitry inside the player reads ahead slightly. If bumped, the section with play from memory without interruption; kind of like time-delay TV.

Digital Output: Some newer CD players have optical and coaxial digital outputs.

Disc-Error Correction: This will help the machine compensate for scratches and imperfections on the compact disc.

DSP: Same as those found in A/V receivers. It simulates acoustic environments.

Headphone Jack with Volume Control: This is great if the CD player is by itself. A receiver does not need to be connected for this feature to work.

Instant-On for Dubbing with a Tape Deck: When you hit the record button of the tape deck, the CD player will turn itself on.

TAPE DECKS

2/3 Head: Some tape decks have two heads, and higher-end models often contain three. If you are going to being doing tons of recording, you need the three heads. They let you monitor the music from the tape as you record.

Auto-Reverse: When the playing tape reaches the end, the other side is automatically played without you having to flip the tape.

Bias Adjustment: This lets you fine-tune the bias for the cassette tape being used. Most machines do this automatically now.

CD-Synchro Record: Starts to record when the CD track is switched on.

Dolby B/C Noise Reduction: The circuitry that eliminates noise from the tapes moving across the audio heads.

Dolby HX Pro: This refines the high-frequency sounds. Sometimes it can actually ruin a dynamic treble section in the music.

Headphone Jack with Volume Control: Same as the CD player's feature.

Memory Stop: The machine will stop when it reaches a preset number.

Microphone Inputs/Controls: Used to plug a microphone into the tape deck. It typically contains a preamp and recording level indicator.

Relay Playback: Both sides of tape A are played, then tape B is played automatically.

DVD PLAYERS

Audio CD/Video CD Compatible: Will play regular audio and video CDs as well as DVD movies.

Built-in Dolby Digital Decoder: A DD-ready receiver does not contain the AC-3 decoder. This means the DVD player has it built in.

Digital Outputs: Contains an optical or coaxial digital output for audio.

Freeze Frame, Frame Advance, Variable Speed Play: These are similar to your VCR's controls. They let you pause the action to a pristine, still picture, click it to the next frame, and adjust the play speed.

Letterbox, Pan & Scan, 16 x 9: Lets you adjust the screen's format to either.

Onscreen Programming/Controls: Same as on an A/V receiver.

NOTE: The proposed DVD dual-sided discs require you to flip them to view the other side. Players will soon contain the mechanics that will read both sides of the disc.

APPENDIX
WEB ADDRESSES

Advent
International Jensen, Inc.
25 Tri-state International Office Center
Suite 400
Lincolnshire, IL 60069
800-477-3257

Aiwa America, Inc.
800 Corporate Dr.
Mahwah, NJ 07430
1-800-BUY-AIWA
http://www.aiwa.com/

Altec Lansing Technologies, Inc.
Milford, PA 18337-0277
800-ALTEC-88
http://www.altecmm.com

B&W Loudspeakers Ltd.
Meadow Road, Worthing
West Sussex, BN11 2RX
Great Britain.
+44 (0) 1903 524801
http://www.bwspeakers.com/

Bose Corporation
Mountain Rd.
Framingham, MA 01701
800-444-2673

Cambridge SoundWorks
1-800-367-4434
http://www.hifi.com/

Denon Electronics Ltd.
Attn: Customer Service
222 New Road
Parsippany NJ 07054
973-575-7810
http://www.denon.com/

Dolby Laboratories
100 Potrero Avenue
San Francisco, CA 94103
1-415-558-0344
http://www.dolby.com/

FCC
http://www.fcc.gov/

GE
Thompson Consumer Electronics Inc.
10330 N. Meridian St.
Indianapolis, IN 46290
800-336-1900
http://www.ge.com/

Harman-Kardon
http://www.harmankardon.com/

Hitachi
3890 Steve Reynolds Blvd.
Norcross, GA 30093
800-448-2244
http://www.hitachi.com

Infinity Systems
http://www.infinitysystems.com

JBL
http://www.jbl.com

JVC America
41 Slater Drive
Elmwood Park, NJ 07407
800-252-5722
http://www.jvc.com/
http://www.jvc-america.com/

Kenwood
Kenwood USA Corp.
P.O. Box 22745
Long Beach, CA 90801-5745
800-536-9663
http://www.kenwoodusa.com/

Linn Products, Ltd.
http://www.linn.co.uk/

Marantz America, Inc.
440 Medinah Road
Roselle, IL 60172
1-800-270-4533
630-307-3100
http://www.marantzamerica.com/

Monster Cable Products, Inc.
274 Wattis Way,
South San Francisco, CA 94080
415-871-6000
http://www.monstercable.com/

NAD Electronics A/S
Grynderupvejen 12,
DK-9610 Nørager
Denmark
+45 9672 1111
http://www.nad.co.uk

Nakamichi America Corp.
955 Francisco Street,
Torrance, CA 90502
310-538-8150
http://www.nakamichiusa.com/

Onkyo
http://www.onkyo.co.jp

Paradigm Electronics, Inc.
905-632-0180
http://www.paradigm.ca/

Philips/Magnavox
Philips Consumer Electronic Company
1-800-531-0039
http://www.philipsmagnavox.com/

Pioneer Electronics, Inc.
Attn: Customer Service
Box 1760
Long Beach, CA 90801
800-746-6337
http://www.pioneer.com/

Polk Audio
5601 Metro Drive
Baltimore, MD 21215
800-377-7655
http://www.polkaudio.com/

Rotel of America
1-800-370-3741
http://mcnaur.com/rotel.html

Samsung
Samsung Electronics America, Inc.
One Samsung Place
Edgewood, NJ 08824
800-767-4675 - #505

Sanyo Fisher Service
21314 Lassen St.
Chatsworth, CA 91311
800-421-5013
http://www.sanyoservice.com/

Sony of America
http://www.sony.com

Sony of Canada Ltd.
405/411 Gordon Baker Road
Willowdale, ONT
Canada, M2H 2S6
416-499-7147
http://www.sony.ca

FAQS AND HELPFUL INTERNET LINKS

Audio Related Internet World Wide Web & FTP Sites
By: Steve Ekblad
http://www.qnx.com/~danh/info.html

Digital Audio Broadcast Info
http://www.kp.dlr.de/DAB
http://www.magi.com/~moted/dr/

DVD Frequently Asked Questions
http://www.nbdig.com/html/faq.htm

MiniDisc Frequently Asked Questions
http://www.hip.atr.co.jp/~eaw/minidisc/minidisc_faq.html

Author's Personal Website
Contains basic electronics information:
http://www.basicelectronics.com
Electronics@pobox.com

Appendix: Web Addresses

Index

SYMBOLS

1-BIT SAMPLING 80
16 X 9 142
16-BIT DAC 106
16-BIT DIGITAL SOUND 108
16-BIT SOUND 106
2/3 HEAD 141
2001: A SPACE ODYSSEY 22
3D IMAGE 38
3D SOUND 8, 33, 107
3D SURROUND SOUND 105
5.1 CHANNELS 16
6-CHANNEL DISCRETE INPUTS 128
8-TIMES OVERSAMPLING 80

A

A/V 17, 41, 42, 50, 66
A/V APPLICATIONS 14
A/V EQUIPMENT 50, 139
A/V JACKS 120
A/V PLAYERS 17
A/V RECEIVER 9, 10, 12, 16, 43, 46, 47, 61, 65-68, 96, 118-120, 126, 127
A/V RECEIVER COMBINATIONS 10
A/V SETUP 9
A/V SIGNAL 65
A/V SPEAKER PACKAGE 96
A/V SPECIALTY STORES 116
ABSORPTION 27
AC-3 44, 47, 139
ACCESSORIES 123
ACOUSTIC 22, 34
ACOUSTIC DIFFRACTION 100
ACOUSTIC EFFECTS 121
ACOUSTIC ENERGY 7, 19, 27, 87, 88
ACOUSTIC ENGINEERING 37
ACOUSTIC ENVIRONMENTS 140
ACOUSTIC PROPERTIES 28, 35, 90
ACOUSTIC SUSPENSION 91
ACOUSTICAL ENERGY 30, 87
ACOUSTICAL PROPERTIES 100
ACOUSTICS 16, 19, 27, 28, 50, 111
ACTIVE SUBWOOFER 49, 96
ACTIVE/PASSIVE SUBWOOFER 140
ADAPTIVE TRANSFORM ACOUSTIC CODING (ATRAC) 83
ADC 79
ADD-ON SOUND CARD 108
ADDRESSES 113
ADJUSTABLE BIAS CONTROL 76
ADS 112
ADVANCED WAVEFFECT (AWE) 106
ADVENT 114
AIR 19, 21-23, 89
AIR MOLECULES 21
AIR PRESSURE 5, 23
AIR VIBRATIONS 5
AIRY 28
AIWA 114
ALBUM 9, 76, 81, 83
ALL-IN-ONE STEREO SYSTEMS 11
ALTERED ENVIRONMENTS 47
ALTERNATING CURRENT 57
AM RADIO STATIONS 63
AM/FM 6, 16
AM/FM DIAL 64
AM/FM SIGNALS 62
AM/FM TUNER 67
AMBIANCE 39
AMBIANCE SWITCH 47
AMBIENCE BUTTON 66
AMBIENT SOUND 39
AMP 12, 14, 48, 56-58, 68, 96, 98
AMP MANUFACTURER 120
AMP/SPEAKER COMBO 111
AMPERAGE 16
AMPLIFICATION 5, 7, 17, 96
AMPLIFICATION COMPONENTS 7
AMPLIFICATION METHOD 56
AMPLIFIED 128
AMPLIFIED SIGNAL 56, 99
AMPLIFIED SOUND SIGNAL 92
AMPLIFIED SPEAKERS 109

AMPLIFIER 7, 11, 15, 16, 19, 29,
    30, 49, 51-53, 55, 57, 67, 89,
    96, 102, 103, 111, 119, 120
AMPLIFIER CHANNEL 34
AMPLIFY 51
AMPLITUDE 16, 22, 24, 29, 37, 52,
    55, 63, 80
AMPLITUDE INFORMATION 38
AMPLITUDE MODULATION (AM) 16, 62
AMPS 49
ANALOG 16, 76
ANALOG AUDIO IN 127
ANALOG COMPONENTS 17
ANALOG COUSINS 75
ANALOG FM 65
ANALOG SIGNAL 78, 80, 106
ANALOG SOUND WAVE 77, 78
ANALOG TAPE RECORDINGS 75
ANALOG VOLTAGE VALUES 79
ANALOG WAVE 78, 84, 105
ANALOG-TO-DIGITAL CONVERTER (ADC)
    78
ANALOG/DIGITAL (A/D) 16
ANTENNA 16, 64
ANTENNA SIGNAL 128
APPLE MACINTOSH COMPUTERS 108
APPLIANCE REPAIRS 118
ATRAC 83
ATRAC 2 83
ATRAC 4.5 83
AUDIBLE LEVELS 7
AUDIENCE 38, 39
AUDIO 5, 9, 14, 16-18, 39, 41, 46, 62,
    72, 75, 76, 112, 114, 123
AUDIO AMPLIFIERS 7, 24, 52
AUDIO APPLICATIONS 63
AUDIO BROADCAST 65
AUDIO BUDGET 117
AUDIO CABLES 124
AUDIO CD 17, 105, 142
AUDIO CHAIN 119
AUDIO CHANNELS 45
AUDIO COMPANIES 85
AUDIO COMPONENTS 22, 53, 69
AUDIO COMPRESSION 83
AUDIO COMPRESSION TECHNOLOGY 65
AUDIO CONSUMER 11, 12
AUDIO DEVICE 123
AUDIO EQUIPMENT 5-8, 11, 17, 24,
    27, 28, 30, 34, 41, 50-52,
    54, 56, 61, 64, 96, 97,
    116, 119, 123
AUDIO FACSIMILES 73
AUDIO FREQUENCIES 62
AUDIO HEADS 72-74, 121
AUDIO I/OS 67
AUDIO INFORMATION 71-73, 83
AUDIO MAGAZINES 103, 113

AUDIO MARKET 8, 9, 84, 114
AUDIO NEEDS 117
AUDIO PATCH CABLES 124
AUDIO PATCH CORD 125
AUDIO PRESENTATION 57
AUDIO PRODUCT 42
AUDIO PRODUCTION 72
AUDIO SIGNAL 5, 88, 89
AUDIO SIGNALS 17, 46, 66
AUDIO SOURCES 61
AUDIO SPECIALTY STORE 117
AUDIO SPECTRUM 115
AUDIO STORES 48, 113
AUDIO SYSTEM 5, 14, 15, 18, 19, 28,
    52, 111, 125
AUDIO SYSTEM ACOUSTICS 28
AUDIO TAPE 36
AUDIO TECHNICA 85
AUDIO TECHNOLOGY 56
AUDIO TERMS 16
AUDIO TUBES 56
AUDIO WAVES 76, 79
AUDIO-ONLY INPUTS 120
AUDIO-RECORDING HEAD 73
AUDIO/VISUAL (A/V) 17, 61, 65
AUDIO/VISUAL (A/V) RECEIVERS 9
AUDIOPHILES 9, 28, 56, 73, 97
AUDIOTAPE 17
AUDITORIUM 38
AUTO REWIND 76
AUTO-REVERSE 121, 141
AUTOMOBILE 52
AVERAGE AUDIO SYSTEM 13
AWE 106

B

B&W 114
BACK SPEAKERS 43
BACK TO THE FUTURE 55
BACKGROUND NOISE 27
BALANCE 7, 53
BALANCE CONTROL 53, 99
BANANA CONNECTORS 126
BAND 39
BASIC AUDIO 45
BASIC SOUND 109
BASS 7, 15, 29, 95, 109
BASS CONTROL 53
BASS DRUM 58
BASS FREQUENCIES 53
BASS REFLEX 91
BASS-HEAVY 111
BASS/TREBLE CONTROLS 67
BASS/TREBLE RESPONSE 114
BASSY 28
BATTERY 102
BEST BUY 116

BI-AMP 140
BI-WIRING 140
BIAS ADJUSTMENT 141
BINARY NUMBER 78, 79
BINARY NUMBER SYSTEM 77, 78
BINARY SIGNAL 78
BINARY VALUE 106
BITS 73, 78, 79
BLACK CONNECTOR 98
BLACK WIRE 98
BLURRED 28
BOARD 108
BODY 26
BOOKS 112
BOOM BOX MODELS 84
BOSE 95, 114
BOX 95
BRAIN 19, 22
BRAND 112-114, 116, 118
BRAND NAME 14, 48
BRANDS 112
BROADCASTERS 46
BROADCASTING 8
BROCHURES 113
BUDGET 16, 119
BUILDING BLOCKS 123
BUILT-IN AMPLIFIERS 69
BUILT-IN AMPS 99
BUILT-IN DECODER 45
BUMP RESISTANCE 121, 141
BUS 26
BUTTON 47, 64, 73
BYTES 77, 78

C

CABINETS 19, 64, 89, 90, 94, 121
CABLE 46, 47, 49, 64, 129
CABLE COMPANIES 46, 47
CABLE MANUFACTURERS 129
CABLE PROVIDER 46
CABLES 11, 123
CALCULATIONS 25
CAMBRIDGE SOUNDWORKS 95, 114
CAPACITORS 5, 68
CAR 107
CAR AUDIO SYSTEM 18
CAR IGNITIONS 63
CAR STEREO 84
CARD 107, 108
CAROUSEL 80
CAROUSEL SYSTEM 9
CARRIER FREQUENCY 64
CARRIER WAVE 62
CARTRIDGE 18, 83-85
CASABLANCA 41
CASSETTE 34, 73
CASSETTE DECK 9, 71-73, 75

CASSETTE TAPE 17, 71, 73, 75, 85
CD 7, 9, 14, 17, 18, 54, 57, 72, 75-85, 99, 115
CD BURNERS 72, 76
CD CHANGER 14, 80, 82, 85, 121, 141
CD FORMAT 14
CD JUKEBOXES 141
CD PLAYER 9-12, 30, 47, 53, 71, 76-78, 80-82, 85,
    93, 100, 106, 114, 119, 121, 131, 141
CD PLAYER HOOKUP 131
CD RECORDING 75
CD SOUND 81
CD SYNCHRO 76
CD TECHNOLOGY 71, 84
CD-AUDIO 105
CD-QUALITY AUDIO BROADCAST 65
CD-ROM DRIVE 101, 105
CD-ROM MARKET 72
CD-ROM/SB/SPEAKER COMBO 108
CD-SYNCHRO RECORD 142
CENTER 67
CENTER CHANNEL 44, 58
CENTER DIALOG SPEAKER 96
CENTER LINE 97, 99
CENTER SPEAKER 44, 48, 94, 96, 99
CENTER STAGE 36
CHANNELS 8, 17-19, 34, 35, 37-39, 43-46, 48, 49, 57, 58, 61, 63, 65, 69, 99, 136
CHARACTERISTICS OF SOUND 27
CHEST 5
CHIP 108
CHIP SET 106, 108
CHROMIUM DIOXIDE 73
CINEMATIC AUDIO 41
CIRCUIT BREAKER 140
CIRCUIT CITY 116
CIRCUITRY 14, 61, 65, 66, 73
CIRCUITS 10, 73
CLASSICAL MUSIC 29
CLOSED SEALS 101
COAXIAL CABLE 124, 126
COAXIAL CONNECTORS 126
COIL 74, 106
COMBINATION COMPONENT SYSTEM 118
COMBINATION HI-FI SYSTEM 12, 14
COMBINATION SYSTEM 10-13, 66, 116, 119
COMMANDS 106
COMPACT CASSETTE 71
COMPACT DISC (CD) 17, 30, 75, 76
COMPACT DISC (CD) PLAYER 6
COMPANIES 10, 11, 95, 106
COMPONENT SETS 10
COMPONENTS 9, 11, 12, 14, 16, 17, 51, 54, 61, 64, 68, 69, 88, 90, 91, 100, 111, 114, 118, 119, 123, 124, 128

Index 151

COMPOSITE VIDEO 47, 127
COMPOSITION 9
COMPRESSION 22
COMPUTER 76, 82, 101, 105-109
COMPUTER CARD 106
COMPUTER CHIP 39
COMPUTER FILES 72
COMPUTER SETUP 105
COMPUTER SOUND 105, 108, 109
COMPUTER SPEAKER MANUFACTURERS 109
COMPUTER SPEAKERS 105
COMPUTER UPGRADE 108
CONCERT ARENA 61
CONCERT HALL 34, 39, 47, 66, 111
CONCERT HALL EXPERIENCE 38
CONCRETE 90
CONDENSER MICROPHONE 102
CONE 7, 18, 23, 55, 89, 93
CONFLICTS 108
CONNECTIONS 15, 101, 126, 127, 129
CONNECTORS 123, 125
CONSTRUCTIVE INTERFERENCE 24
CONSUMER 65, 76, 118
CONSUMER A/V MARKET 47
CONSUMER AUDIO EQUIPMENT 72
CONSUMER ELECTRONICS 18
CONSUMER ELECTRONICS DEVICE 76
CONSUMER MAGAZINE 103, 113
CONSUMER MARKET 47
CONSUMER REPORTS 114
CONSUMER SPEAKERS 48
CONSUMERISM 114
CONTROL AMPLIFIER 53, 54
CONTROL SECTION 53
CONTROLS 11, 14, 15, 18, 53, 54, 66, 68, 69, 118, 120
CONVENTIONAL SPEAKERS 94
CONVERTERS 105
COPY 75
CORD 99, 101-103
CORDLESS HEADPHONES 101
CORROSION 129
CORROSIVE ELEMENTS 129
CORROSIVE SQUARE WAVE 57
COST 114
CRISP 28
CROSSOVER NETWORK 90, 92, 93
CURRENT 57, 68
CYCLES PER SECOND 23, 25
CYMBAL 58

D

DAB 17, 65
DAB STANDARDS 65
DAC 78, 79, 80, 105, 106
DAT (DIGITAL AUDIO TAPE) 9, 14, 16, 17, 75, 76, 85, 120, 128
DAT DECKS 76
DAT MERCHANDISE 76
DATA REDUCTION SCHEME 44
DBS 44, 46, 47
DBS EQUIPMENT 47
DD 66
DD DECODER 48
DD DEVICE 48
DD SIGNAL 48
DD-READY RECEIVERS 69
DECIBEL INFORMATION 37
DECIBELS (DBS) 25, 26, 29, 30, 36
DECIMAL NUMBER 77
DECK 73, 84
DECODER 9, 11, 13, 18, 43, 45-48, 61, 68
DECODER HARDWARE 61
DECODES 39
DECODING SCHEME 18
DEDICATED SUBWOOFER SIGNAL 44
DEGRADATION 53
DEMODULATION 62
DENON 85, 114
DEPARTMENT STORES 116
DESIGN 11, 89
DESIGNERS 89, 94
DESTRUCTIVE INTERFERENCE 24
DETACHABLE SPEAKERS 11, 118
DETAILED 29
DEVICE 120
DIALOG CHANNEL 44
DIALS 14, 64
DIFFRACTION 100
DIGITAL 17, 47, 65, 76-78, 84, 105, 106
DIGITAL AUDIO BROADCAST (DAB) 17, 65
DIGITAL AUDIO IN 128
DIGITAL AUDIO SIGNALS 124
DIGITAL AUDIO TAPE (DAT) 75
DIGITAL AUDIO TAPE DECK 11
DIGITAL AUDIO/VIDEO RECEIVER 65
DIGITAL AUDIO/VISUAL TECHNOLOGY 69
DIGITAL CABLES 124
DIGITAL CASSETTE 75
DIGITAL CONNECTORS 126
DIGITAL DATA 9, 75
DIGITAL DECODING CIRCUITRY 69
DIGITAL DEVICE 76
DIGITAL DISC 75
DIGITAL ELECTRONICS 17
DIGITAL FM 65
DIGITAL HEADPHONES 101
DIGITAL I/OS 67
DIGITAL INFORMATION 78, 79
DIGITAL INPUTS 120
DIGITAL INSTRUMENT 106
DIGITAL MIDI CAPABILITIES 108

DIGITAL OUTPUT 48, 99, 141, 142
DIGITAL PLAYER DEVICES 44
DIGITAL RECEIVERS 48
DIGITAL SATELLITE AUDIO 128
DIGITAL SATELLITES 46
DIGITAL SIGNAL 65, 80, 99, 124
DIGITAL SIGNAL PATCH CORDS 126
DIGITAL SIGNAL PROCESSOR (DSP) 61, 66, 99
DIGITAL SOUND 106, 111
DIGITAL SOUND EFFECTS 105
DIGITAL SPEAKER SYSTEM (DSS) 99
DIGITAL TECHNOLOGY 106
DIGITAL TELEVISION 17, 65, 128
DIGITAL TELEVISION BROADCAST STANDARDS 47
DIGITAL TELEVISION STANDARDS 44
DIGITAL THEATER SYSTEMS (DTS) 49
DIGITAL TV STATIONS 44
DIGITAL VIDEO DISC (DVD) 9, 17, 71, 82
DIGITAL VIDEO DISC PLAYERS 47
DIGITAL-QUALITY MUSIC 46
DIGITAL-TO-ANALOG CONVERTER (DAC) 78, 105
DIGITALLY 17
DIGITIZED INSTRUMENT SOUNDS 106
DIMENSION 19
DIRECT BROADCAST SATELLITE (DBS) 44
DIRECT SOUND 39
DIRECTION 36
DIRECTIONAL DRIVER 98
DIRECTIONAL INFORMATION 39
DIRECTIONAL MICROPHONE 102
DIRECTOR 50
DISC 18, 71, 76, 78, 79, 83, 84
DISC-ERROR CORRECTION 141
DISCRETE CHANNELS 16
DISTANCE 25, 31, 98
DISTANCE INFORMATION 38
DISTORTION 8, 15, 18, 26, 30, 55, 57, 75, 81, 95, 99, 101, 115
DISTURBANCE 29
DNR 75, 76
DOLBY 10, 12, 18, 39, 44, 49, 50, 72, 73
DOLBY 3 STEREO MODE 68, 139
DOLBY AC3 44
DOLBY B 121, 142
DOLBY C 121, 142
DOLBY CIRCUITRY 45, 69
DOLBY DECODING CIRCUITRY 47
DOLBY DIGITAL 8, 13, 16-18, 39, 44, 47, 48, 50, 58, 66, 68, 96, 119, 120
DOLBY DIGITAL AUDIO SIGNAL 135
DOLBY DIGITAL CIRCUITRY 47
DOLBY DIGITAL DECODER 139, 142
DOLBY DIGITAL DECODING CIRCUITRY 135
DOLBY DIGITAL READY 48
DOLBY DIGITAL RECEIVER 44, 68, 69, 128
DOLBY DIGITAL SIGNALS 44
DOLBY DIGITAL SOUNDTRACK 47
DOLBY DIGITAL SOURCES 47
DOLBY DIGITAL SURROUND SOUND 47
DOLBY DIGITAL-READY RECEIVER 128, 135
DOLBY DIGITAL-READY UNIT 120
DOLBY HX PRO 142
DOLBY LABORATORIES 18, 43, 75
DOLBY LOGO 122
DOLBY NOISE REDUCTION (DNR) 75, 121, 142
DOLBY NOISE REDUCTION CIRCUITRY 14, 73
DOLBY PRO LOGIC 13, 16, 18, 43, 46, 47, 50, 66, 99, 119, 122
DOLBY PRO LOGIC CIRCUITRY 139
DOLBY PRO LOGIC DECODER 45
DOLBY PRO LOGIC RECEIVER 11, 120
DOLBY PRO LOGIC SIGNAL 46
DOLBY PRO LOGIC SOURCES 46
DOLBY SURROUND SIGNAL 46, 61
DOLBY SURROUND SOUND 18, 41, 43, 49
DOLBY SURROUND SOUND DECODER 43
DOLBY SURROUND SOUND MOVIE 45
DOLBY SURROUND SOUND SIGNAL 65
DOLBY SYSTEM 96
DOLBY TECHNOLOGY 75
DOLBY-EQUIPPED AUDIO PRODUCTS 45
DOMES 18
DOMINANT SOUNDS 83
DOPPLER EFFECT 107
DRIVERS 7, 18, 19, 23, 26, 89-93, 95, 96, 120, 121
DRUM 21, 27, 29
DRUM SKIN 21
DSP 66, 99, 140, 141
DSS 99
DSS LINK 100
DSS SPEAKERS 100
DTV 44, 47, 48, 120
DUAL DECK 74, 76, 121
DUAL STEREO SIGNAL 63
DUAL-CASSETTE DECK 85
DUAL-DECK UNIT 14
DUAL-TAPE DECK 119
DUBBING 75, 141
DUPLICATES 72
DVD 10, 16, 44, 47, 48, 71, 81, 85, 120, 135

Index 153

DVD DISCS 17
DVD PLAYER 9, 11, 14, 16, 17, 44, 48, 50, 71, 82, 126, 128, 135, 142
DVD PLAYER HOOKUP 135
DVD-AUDIO 71, 82, 85
DYNAMIC MICROPHONES 102
DYNAMIC SPEAKERS 89

E

E-MAIL 105, 112
EARDRUMS 5, 19, 22
EARS 5, 26, 51, 93, 100
ECONOMY PRODUCTS 116
EDITING 43
EFFECTS 16
ELECTRIC MOTORS 63
ELECTRIC SIGNALS 84
ELECTRICAL DEVICES 73
ELECTRICAL ENERGY 7, 22, 27, 29, 52, 87, 102
ELECTRICAL IMPULSES 30, 87
ELECTRICAL NOISE 57
ELECTRICAL POWER 55
ELECTRICAL SIGNAL 16, 19, 34, 51
ELECTRICAL SOUND IMPULSES 30
ELECTRICAL WAVES 87
ELECTRICITY 17, 52, 77, 102
ELECTRODYNAMIC 89
ELECTRODYNAMIC PRINCIPLES 88
ELECTRODYNAMIC SPEAKERS 7
ELECTRODYNAMICS 18, 29
ELECTROMAGNET 7, 30, 74, 89
ELECTROMAGNETIC 84
ELECTROMAGNETIC COIL 93
ELECTROMAGNETIC DEVICES 27
ELECTROMAGNETIC RADIO WAVES 16, 61
ELECTROMAGNETS 88
ELECTRONIC AMPLIFIERS 7
ELECTRONIC AUDIO EQUIPMENT 33
ELECTRONIC AUDIO SYSTEMS 27
ELECTRONIC BOUTIQUES 118
ELECTRONIC CIRCUIT 63
ELECTRONIC COMPONENT 17, 50, 55, 76
ELECTRONIC ENERGY 19
ELECTRONIC SIGNAL 52, 55
ELECTRONIC SYSTEMS 27
ELECTRONIC WAVES 31
ELECTRONICS 27, 31, 77, 88, 96, 107, 113
ELECTRONICS BOUTIQUES 116, 117
ELECTRONICS JARGON 117
ELECTRONICS STORES 122
ELECTRONS 5, 31, 52
ELECTROSTATIC SPEAKERS 89
ELEMENTARY SYSTEMS 14
EMERGENCY SOUNDS 101

EMOTIONS 29
ENCLOSURES 11, 91, 94
ENCODED SIGNAL INFORMATION 45
ENCODED SIGNALS 10
ENCODES 39
ENCOMPASSING SOUND 18
END TABLE 19
ENERGY 6, 19, 21, 22, 29, 30, 52, 87
ENERGY FORM 27
ENERGY TRANSFER 22
ENGINEERING 90
ENGINEERS 34, 38, 50, 66, 83
ENTERTAINMENT 5
ENTERTAINMENT CENTER 95
ENTERTAINMENT SYSTEM 67
ENVIRONMENT 16
EQUALIZATION 7, 53
EQUALIZER 67
EQUILATERAL TRIANGLE 97
EQUIPMENT 12, 27, 31, 36, 45, 67, 72, 97, 129
EQUIPMENT LIST 119
EXPANDED AUDIO SYSTEM 14
EXPANSION ABILITY 48
EXPLOSION 24
EXTENDED-FREQUENCY HEARING 27
EXTERNAL AMPLIFIERS 128, 136
EXTERNAL AMPLIFIERS HOOKUP 136
EXTERNAL NOISE 10, 100, 103

F

F-CONNECTORS 126
FACE 5, 26
FEATURES 112, 114, 139
FERRIC OXIDE 73
FIBER OPTICS 124
FIBERBOARD 90
FIDELITY 8, 102
FILM TRACK 43
FILMING 43
FINE TUNE 64
FLIERS 112
FLOOR PLAN 96, 97
FLUTTER 121
FM 63, 106
FM ANTENNA INPUTS 64
FM RADIO 65
FM RADIO BAND 63
FM RECEPTION 114
FM SCREWS 64
FM SIGNALS 63, 64
FM WAVES 63, 64
FM/MIDI PLAYBACK 108
FM/RF 128
FORMULA 25
FRAME ADVANCE 142
FREEZE FRAME 142

FREQUENCY 18, 19, 22, 23, 25-29, 36, 37, 61, 64, 80, 100, 107
FREQUENCY BAND 15, 16, 53, 62, 63, 97
FREQUENCY MODULATION (FM) 16, 63, 106
FREQUENCY RANGE 8, 18, 26, 29, 92
FREQUENCY RESPONSE 29
FREQUENCY/AMPLITUDE INFORMATION 38
FRONT A/V INPUTS 140
FRONT SET 94
FRONT SPEAKERS 43, 48, 96, 99
FRONT STEREO PAIR 96
FRONT WAVE 91
FURNITURE 10
FURNITURE PLACEMENT 97
FUSE 140

G

GADGETS 14, 139
GE 116
GEARS 72, 73
GENERAL MIDI (GM) 106
GEOMETRY 15
GLASS 124
GM 106
GOLD CONNECTIONS 16, 129
GOLD PLATING 129
GROOVE 84
GUIDELINE 123
GUITAR 29

H

HARDWARE 108
HARMAN-KARDON 114
HEADPHONE JACK 141, 142
HEADPHONE VOLUME CONTROLS 121
HEADPHONES 10, 39, 87, 100, 101
HEADS 73, 75
HEAR 26
HEARING RANGE 26
HEAVY METAL 55
HEAVY WIRE 96
HERTZ (HZ) 23
HETERODYNE RECEIVER 63
HI-FI 7, 8, 12, 22, 38, 50, 71, 85, 105
HI-FI AUDIO SYSTEM 5
HI-FI COMPONENTS 10
HI-FI EQUIPMENT 26, 33, 36
HI-FI RECORDING 9
HI-FI STEREO EQUIPMENT 9
HI-FI SYSTEM 8, 14
HI-FI UNITS 42
HIGH CURRENTS 68
HIGH FIDELITY (HI-FI) 7, 8, 71
HIGH FREQUENCIES 28, 29
HIGH-FIDELITY AUDIO SYSTEM 26

HIGH-FREQUENCY DRIVER 92
HIGH-PRESSURE AREAS 21
HIGH-PRESSURE COMPRESSION 22
HIGH-QUALITY A/V SIGNAL 65
HIGHS 53
HOME 50, 113
HOME DECK 18, 84
HOME ENTERTAINMENT CENTER 126
HOME ENTERTAINMENT SYSTEMS 109
HOME MARKET 43
HOME STEREO 100
HOME STEREO SYSTEM 109
HOME STEREO SYSTEMS 9, 109
HOME THEATER 9, 11, 12, 16, 18, 41, 47, 48, 66, 82, 111
HOME THEATER COMPONENTS 54
HOME THEATER MARKET 43
HOME THEATER RECEIVERS 17, 122
HOME THEATER SOUND SYSTEMS 12, 41
HOME THEATER STEREO 41
HOME THEATER SURROUND SOUND 13
HOOKUP 47, 123, 125
HOOKUP DEVICES 123
HOOKUP TECHNIQUES 123
HOT 29
HUMAN EAR 25, 26
HUMAN HEARING 23, 25, 80
HUMAN HEARING FREQUENCY RANGE 8

I

IBM-COMPATIBLE MACHINES 108
IMPEDANCE 15, 19, 57, 58, 89, 99
IMPEDANCE LEVEL 99
IMPEDANCE MATCHING 57
IMPULSE 29
INAUDIBLE RANGE 80
INCOMING SIGNAL 64
INDEPENDENT SOURCE 17
INDY CAR 107
INFINITY 114
INFORMATION 17, 46, 77
INFRARED LIGHT SIGNALS 66
INPUT AUDIO WAVE 79
INPUT LIMITS 120
INPUT SELECTOR 140
INPUT SIGNAL 54, 92
INPUT SOUND SIGNAL 80
INPUT SOURCES 47
INPUTS 36, 67, 128
INSTANT-ON 141
INSTRUCTION 66
INSTRUMENT MIXING 115
INSTRUMENT SOUNDS 106
INSTRUMENTS 29, 35, 98, 106
INTEGRATED SPEAKERS 11
INTEGRATION 69
INTENSITY STEREO 36

Index 155

INTERFERENCE 24, 56, 91
INTERNALLY-POWERED ACTIVE
    SUBWOOFER 49
INTERNET 113

J

JARGON 112
JAZZ CLUB 47
JAZZ HALL 61
JBL 95
JBL CONSUMER 114
JBL PRO 114
JBL SPEAKERS 11
JET 107
JUNCTION 129
JURASSIC PARK 96
JVC 114, 123

K

KETTLE WHISTLING 100
KINETIC ENERGY 22
K-MART 114, 116
KNOBS 14

L

LANDS 76, 77
LARGE TELEVISION 18
LASER 77, 79
LASER BEAM 76
LASER LIGHT 76
LASER VIDEO DISC 47
LASERDISC 47
LASERDISC PLAYER HOOKUP 135
LASERDISC PLAYERS 44, 47, 135
LAW OF THE FIRST WAVEFRONT 36
LAYOUT 97
LED DISPLAY 140
LEFT 67
LEFT CHANNEL 43, 44, 45, 57, 58
LEFT FREQUENCY 38
LEFT LEVELS 36
LEFT SURROUND CHANNEL 44
LEFT TRACK 36
LETTERBOX 142
LETTERS 76
LEVEL 52, 53, 58, 111
LEVEL CONTROL 53
LEVERS 53
LIGHT 24, 77
LIGHT AUDIO SYSTEM 14
LIGHT SWITCH 77
LIGHTNING STORM 24
LINE-LEVEL 127
LINE-LEVEL OUT 128
LINN 114
LISTENER 33, 34, 38, 39, 54

LISTENING AREA 16
LISTENING LEVELS 49
LISTENING ROOM DESIGN 27
LIVE RECORDING 39
LIVE SIGNAL 6
LIVING ROOM 10, 18, 41, 50
LO-FI 8
LOADING MECHANISMS 73
LOCAL ADS 117
LOGO 46
LOUDNESS 25, 26, 29
LOUDNESS LEVEL 99
LOUDSPEAKER CABINETS 89, 90
LOUDSPEAKER CONFIGURATIONS 10
LOUDSPEAKER DRIVERS 15
LOUDSPEAKERS 10, 18, 19, 26, 39, 42,
    51, 52, 54, 87-95, 114, 120, 140
LOW FREQUENCIES 26, 28, 92
LOW FREQUENCY DRIVER 92
LOW-END MARKET 14
LOW-FIDELITY 63
LOW-PRESSURE AREAS 21
LOW-PRESSURE RAREFACTION SECTION
    22
LOWS 53
LP 9, 10, 11, 14, 30, 51, 71, 81, 84, 85
LP NEEDLE 52
LP RECORD PLAYERS 85

M

MACHINE 79
MACINTOSH COMPUTERS 108
MAGNET 7, 49, 89
MAGNETIC COIL 93
MAGNETIC FIELD 74, 93
MAGNETIC MATERIAL 74
MAGNETIC PARTICLE 74
MAGNETIC SHIELDING 109, 121, 141
MAGNETIC SIGNATURE 74
MAGNETICALLY-CHARGED PARTICLES
    71, 73
MAGNETO-OPTICAL (MO) 83
MANN'S CHINESE THEATER 50
MANUFACTURER MANUALS 123
MANUFACTURERS 10, 12, 18, 42, 47, 48,
    58, 84, 95, 109, 112-114
MARANTZ 114
MARANTZ RECEIVER 11
MARBLE 90
MASH 80
MASS-MARKET RECEIVERS 56
MASS-MARKET UNIT 58
MASS-MARKETERS 116
MATCHBOX CASSETTE 75
MATCHING IMPEDANCE 57
MATERIAL 90, 125
MATTER 22

MD 83
MEASUREMENTS 25, 29, 80
MECHANICAL AMP 55
MECHANICAL DEVICES 73
MECHANICAL ENERGY 27
MECHANICAL MOTION 84
MEDIUM 22
MEDIUM FREQUENCY DRIVER 92
MEMORY STOP 142
METAL 73
METAL-COATED TAPE 73
METALLIC PARTICLES 74
MICROPHONE CONTROLS 142
MICROPHONE DIRECTIONALITY 102
MICROPHONE HOOKUP 102
MICROPHONE INPUTS 76, 142
MICROPHONE PREAMP 54
MICROPHONES 6-8, 10, 19, 29, 30, 34, 36-38, 51-54, 87, 89, 101, 102
MICROPROCESSORS 5, 17, 61, 66, 77
MIDI 105, 106, 108
MIDI CAPABILITIES 108
MIDI COMPATIBLE 106
MIDRANGE 16, 18, 29, 92, 94, 95, 115
MIDRANGE DRIVERS 26
MINI-SPEAKERS 14
MINI-THEATER AUDIENCE 42
MINIDISC 9, 10, 14, 16, 18, 71, 72, 76, 82-85, 120, 128
MINIDISC RECORDERS 84
MIXING 9, 54, 75
MIXING STAGE 50
MO 83
MODEL 9, 112, 113, 118
MODERN MEDIA 31
MODES 140
MODULAR 11
MODULATED FREQUENCY 63, 64
MODULATION 62
MOLECULES 22
MONAURAL 34
MONITOR 18, 109, 128
MONO 19, 44, 82, 109
MONO SOUNDTRACK 46
MONO SYSTEMS 34
MONOPHONIC 8, 34
MONOPHONIC SYSTEM 36
MONSTER CABLE PRODUCTS, INC. 136
MOTHERBOARD 106, 108
MOTORS 72, 73
MOVIE 9, 17, 41, 44, 45, 50, 55
MOVIE DIRECTOR 41
MOVIE EXPERIENCE 41
MOVIE SOUND EFFECTS 96
MOVIE SOUNDTRACK 43
MOVIE STAGES 65
MOVIE THEATER 41, 50
MOVIE TRACKS 43-46, 49, 120

MTS STEREO SIGNAL 46
MULTICHANNEL 33
MULTICHANNEL AUDIO SYSTEM 8
MULTICHANNEL SURROUND SOUND SYSTEMS 49
MULTICHANNEL SYSTEM 19, 39
MULTICHANNELING 34
MULTIDIMENSIONAL SOUND 38
MULTIDISC CD 16
MULTIFREQUENCY WAVES 26
MULTIPLE CHANNELS 34, 43
MULTIPLE INPUT SOURCES 47
MULTIPLE MICROPHONES 102
MULTIPLE SPEAKERS 34
MUSIC 9, 10, 26, 29, 39, 54, 58, 66, 71, 79, 83, 94, 98, 100, 111, 120
MUSIC INDUSTRY 76
MUSICAL INSTRUMENT DIGITAL INTERFACE (MIDI) 106
MUSICIANS 19, 66, 98

N

NAD 55, 114
NAKAMICHI 114
NEEDLE 84
NEGATIVE 80, 93
NERVES 5
NETWORK 55
NEWSCAST 71
NHT 114
NOISE 17, 18, 24, 30, 55, 57, 58, 63, 64, 72, 73, 75, 100
NOISE ELIMINATION CIRCUITRY 74
NONCORROSIVE CONNECTION 125
NONDIRECTIONAL 96
NORTH AMERICA 75
NOSTALGIA COLLECTORS 71
NOSTALGIC-SOUNDING EQUIPMENT 56
NUCLEAR ENGINEER 68
NUMBERS 76, 78, 79

O

OBJECT 23
OCEAN WAVE 22
OCEANS 22
OHMS 19, 57
OMNIDIRECTIONAL 96
OMNIDIRECTIONAL MICROPHONE 102
ONE-CHANNEL AUDIO SYSTEM 34
ONES 77
ONKYO 114
ONSCREEN CONTROL 120, 143
ONSCREEN DISPLAY 140
ONSCREEN PROGRAMMING 143
OPEN SEALS 100
OPEN SPACES 38

OPTICAL CONNECTORS 126
ORIGINAL ANALOG SOUND 79
ORIGINAL RECORDING 19, 33, 75
OUTLETS 140
OUTPUT 5, 7, 30, 47, 53, 54, 55, 61, 67, 109
OUTPUT AMPLIFIER 7, 52
OUTPUT PATH 17
OUTPUT POWER 109
OUTPUT SIGNAL 18
OUTPUT STAGE 88
OVERDRIVING 57
OVERSAMPLING 80

P

PACKAGE 48, 119
PACKAGE DEAL 94
PAN & SCAN 142
PARADIGM 114
PARTICLE BOARD 90
PARTS 123
PASSIVE SUBWOOFERS 49, 96
PERMANENT MAGNET 30, 88, 93
PHANTOM POWER 54
PHASE 24, 34
PHONE NUMBERS 113
PHONO 133
PHONO CARTRIDGE INPUT 54
PHONOGRAPH 54, 67, 71, 84
PHONOGRAPH CARTRIDGE 53
PHONOGRAPH INPUT 85
PHOTOCELL 76
PHYSICAL MOVEMENT 87
PIANO 37
PICTURE 16, 18, 63, 76, 82
PICTURE TUBE 49, 121
PIEZOELECTRIC DEVICE 84
PIONEER 55
PISTON 89
PITCH 23, 107
PITS 76, 77
PLASTIC 95
PLASTIC FIBERS 124
PLASTIC TAPE 71, 73, 74
PLAY-AND-RECORD TECHNOLOGIES 72
PLAYBACK 29, 34, 36, 37, 74, 79, 121
PLAYBACK EQUIPMENT 8, 73
PLAYBACK EXPERIENCE 38
PLAYBACK SYSTEM 39
PLAYER QUALITY 114
PLAYERS 75, 80
POLARITY 15, 57, 80, 98
POLK 114
PORT 91
PORTABLE PLAYER 18
PORTABLE RECORDERS 73
PORTABLES 84

POSITIVE 80, 93
POWER 16, 19, 25, 54, 55, 57, 69, 96, 114
POWER AMPLIFIER 7, 13, 18, 47, 52, 54-57, 61, 67, 74, 93
POWER LEVEL 26, 55
POWER LOADS 56, 125
POWER OUTLET 96
POWER RATING 55, 57, 99
POWER SOURCE 102
POWER-UP SEQUENCE 54
PREAMP 11, 47, 53, 54, 56, 67
PREAMPLIFIED OUTPUTS 128, 136
PREAMPLIFIER 7, 13, 18, 29, 52-54, 61, 64, 85, 102
PREDESIGNATED INSTRUMENTS 106
PRELIMINARY AMPLIFICATION 53
PRERECORDED DISC 83
PRERECORDED MUSIC 14, 85
PRERECORDED SOUNDS 30
PRESETS 67, 140
PRESSURE 22, 30
PRESSURE LEVEL 34
PRESSURE WAVES 19, 93
PRESSURE ZONES 22
PRICE GUARANTEE 118
PRO LOGIC 18, 44, 46, 47, 50, 58, 66-68
PRO LOGIC DECODER 48
PRO LOGIC SIGNAL 44, 46
PRO LOGIC SURROUND SOUND 39
PRODUCT 118
PRODUCT INFORMATION 113
PRODUCTION 16, 45
PROFESSIONAL RECORDINGS 102
PROGRAMS 109
PURCHASER 118
PURCHASING DECISIONS 12
PURCHASING STRATEGIES 9, 117
PUSH-BUTTON CONNECTORS 126

Q

QUALITY 83, 121
QUALITY RECORDING 27
QUALITY SPEAKERS 16
QUANTIZATION 80
QUANTIZATION NOISE 80
QUESTIONS 111, 112
QUINTUPLET SPEAKERS 119

R

RACK 97
RACK STEREO SYSTEMS 9
RACK SYSTEM 10-14, 116, 119
RACK SYSTEM RECEIVERS 66
RADIO 71
RADIO BROADCASTING 17

RADIO DRAMA 71
RADIO FREQUENCY 62, 63
RADIO HEADPHONES 101
RADIO RECORDING 8
RADIO STATION 17, 19, 62, 64, 65
RADIO STATION PRESETS 140
RADIO WAVES 61, 62, 64
RANDOM SOUNDS 38
RANGE 101
RAREFACTION 22, 30
RAREFACTION WAVES 22
RATINGS 112, 114
RCA CONNECTORS 124, 125, 126
RCA-TYPE CONNECTOR 124, 127
RE-RECORDABLE DISC 82
REAL-WORLD SOUNDS 101
REAR WAVE 91
RECEIPT POINT 22
RECEIVER 9, 11-13, 18, 19, 43, 45, 47-50, 54, 57, 58, 61, 63, 64, 66-69, 74, 96, 99, 102, 114, 120, 126, 128, 131, 133, 135, 139
RECEIVER FACES 53
RECEIVER MARKET 65
RECEIVER PANEL 68
RECEIVER SIGNALS 54
RECEIVER/DECODER 12
RECEPTION 16
RECHARGEABLE HEADPHONES 101
RECHARGING BASE 101
RECOMMENDATIONS 118
RECORD 14, 54
RECORD COMPANY 76
RECORD LABELS 76
RECORD NEEDLE 51
RECORD PLAYER 84
RECORDABLE DISCS 83
RECORDER 38
RECORDING 8, 9, 34, 36, 38, 39, 43, 73-75, 84, 101, 111, 115, 121
RECORDING ARTIST 33
RECORDING CHANNEL 34, 36
RECORDING DECKS 14
RECORDING DEVICES 30
RECORDING QUALITY 114
RECORDING SOURCES 8
RECORDING TRACK 75
RECORDINGS 19, 34, 106
RECORDS 9, 71, 84
RED CONNECTOR 98
RED WIRE 98
REEL-TO-REEL 73
REFERENCE LEVEL 25
RELAY PLAYBACK 142
REMOTE 66, 118, 126
REMOTE CONTROL 10, 11, 14, 15, 48, 66, 67
REPRODUCING SOUND 8

RESEARCH 112
RESISTANCE 16, 19, 57
RESISTORS 5, 58
RETAIL STORE 113
REVERBERATION 38, 39
REVIEWS 103
REWRITABLE CD 72, 76
REWRITABLE DRIVES 82
RF CABLES 126
RF CONNECTIONS 133
RF TELEVISION SIGNAL 124
RIBBON CABLE 125
RIGHT 67
RIGHT CHANNEL 37, 43, 44, 45, 57, 58
RIGHT FREQUENCY 38
RIGHT LEVELS 36
RIGHT SURROUND CHANNEL 44
RIGHT TRACK 36
ROCK 29
ROMEX 125
ROOF ANTENNA 63
ROOFS 64
ROOM 28, 35, 36, 39, 97, 99
ROTEL 114
ROUTING SYSTEM 54

S

S-VIDEO 47, 127
S-VIDEO CONNECTIONS 120, 126, 128
S/N (SIGNAL-TO-NOISE RATIO) 18
SALABLE PRODUCTS 50
SALES REPRESENTATIVE 122
SALESPEOPLE 28, 113, 116, 117
SALESPERSON 112, 117
SAMPLING 79, 80
SAMPLING RATE 106
SANYO 116
SAT/SUB COMBO 91
SATELLITE ENCLOSERS 95
SATELLITE SPEAKERS 68, 95
SATELLITES 48, 94, 95, 97
SB 108
SB 16 COMPATIBLES 108
SB 32 108
SB 64 108
SB COMPATIBLE 108
SCAN FEATURE 67
SCIENCE 16
SCIENCE OF SOUND 27, 30
SEALS 100
SEMI-CLOSED SEALS 100
SEMI-OPEN SEALS 100
SENSORS 73
SEPARATE AUDIO SYSTEM 14
SEPARATE COMPONENT SYSTEM 10, 12, 119
SEPARATE COMPONENTS 11

Index   **159**

SEPARATE SYSTEM 13
SERVICE 116, 117
SERVICE CONTRACTS 118
SERVICE RECORD 114
SETUP 120
SHELF STEREOS 116
SHIELDED SPEAKER 48
SHIELDING 49
SHURE 85
SIDE EFFECTS 80
SIDE SPEAKERS 109
SIGNAL 6, 7, 8, 16, 19, 30, 43, 52, 53, 55, 61, 62, 64, 74, 84, 93, 99, 101, 102
SIGNAL AMPLIFIER 64, 73
SIGNAL LEVEL 52, 53
SIGNAL PROCESSORS 18
SIGNAL SOURCE 30, 45, 47, 53
SIGNAL SOURCE CONTROL 53
SIGNAL SPLITTER 64
SIGNAL WIRE 96
SIGNAL-TO-NOISE RATIO (SNR OR S/N) 30, 55, 72, 75, 79
SINGER 19, 29, 36
SINGLE PLAYER 80
SKIN 21
SLUSH FUND 118
SNARE DRUM 5, 21
SNR (SIGNAL-TO-NOISE RATIO) 18
SOFTWARE 105, 106, 109
SOLID-STATE TRANSISTOR AMP 56
SOLID-STATE TRANSISTORS 55
SONIC ILLUSION 39
SONIC IMPACT 29
SONIC SPACE 19
SONY 114
SONY CD PLAYER 11
SONY DYNAMIC DIGITAL SOUND (SDDS) 49
SOUND 5-8, 16, 19, 21, 22, 25, 26, 29, 30, 37, 48, 51, 55, 83, 88, 100, 105, 107, 113, 114
SOUND BITS 124
SOUND BLASTER (SB) 108
SOUND BLASTER 16 108
SOUND BLASTER 32/64 AWE 106, 108
SOUND BLASTER PRO 108
SOUND CAPABILITIES 108
SOUND CARD 105-109
SOUND DRIVERS 11
SOUND EFFECTS 96, 105, 106
SOUND ELEMENTS 27
SOUND ENERGY 29, 52
SOUND ENGINEER 50
SOUND EQUIPMENT 5
SOUND EXPANSION CAPABILITIES 108
SOUND FIELD 39, 41
SOUND FREQUENCIES 63
SOUND FREQUENCY 38
SOUND INFORMATION 19, 44, 62, 63, 69
SOUND LANGUAGE 28
SOUND LEVELS 25, 36
SOUND PACKAGE 18
SOUND PRESSURE 25, 36
SOUND PRESSURE LEVEL 25, 26
SOUND PRESSURE WAVES 31, 102
SOUND PRINCIPLES 27
SOUND PRODUCTION 27
SOUND PROGRAMS 108
SOUND QUALITY 15, 50, 82, 98, 101, 105
SOUND RECEPTION 27
SOUND RECORDINGS 27
SOUND SCENARIOS 66
SOUND SIGNAL 7, 12, 17, 24, 55, 64, 74, 92
SOUND SOURCE 36
SOUND STAGE 19, 33, 34, 38, 39, 42, 43, 44, 65, 66, 115
SOUND STORAGE 30, 71, 73
SOUND SYSTEM 41, 108
SOUND TASTES 27
SOUND TECHNOLOGY 105, 107
SOUND THEORY 5
SOUND TRANSMISSION 27
SOUND VIBRATIONS 8
SOUND WAVE 16, 21, 22, 24, 25, 29, 31, 42, 76, 78, 87-89, 106
SOUND WAVE AMPLITUDE 106
SOUND WAVE REFLECTIONS 27
SOUNDPROOF 111
SOUNDPROOF A/V ROOM 16
SOUNDS 27, 33, 34, 38, 57, 71, 84, 102, 107
SOUNDS CARDS 108
SOUNDTRACK 42, 43, 46, 50
SOURCE 5, 6, 34
SOURCE ENERGY 22
SOURCE POINT 22, 31
SPACE 22, 35, 46
SPACE SHIPS 22
SPADE CONNECTOR 126
SPARK 63
SPATIAL IMAGING CAPABILITIES 39
SPATIAL INFORMATION 38, 50, 107
SPATIAL TEXTURE 34
SPEAKER 7-9, 11, 12, 14-16, 18, 19, 27, 29, 30, 31, 34-36, 38, 39, 41, 43-45, 47-50, 52, 53, 55, 57, 58, 66, 68, 74, 79-91, 96-100, 102, 105-108, 111, 112, 115, 119, 122, 125, 126, 130
SPEAKER BOXES 10
SPEAKER CABINET 91
SPEAKER CONE 52
SPEAKER CONFIGURATION 44, 93
SPEAKER DESIGNERS 94

SPEAKER DRIVERS 54
SPEAKER HEIGHT 99
SPEAKER HOOKUP 130
SPEAKER MATCHING 57
SPEAKER OPTION 109
SPEAKER OUTPUTS 48
SPEAKER PACKAGE 14
SPEAKER SELECTIONS 48
SPEAKER SET 120
SPEAKER WIRE 125
SPEAKER WIRING 98, 99
SPEAKERS OUT 128
SPECTRUMS 93
SPEED 25, 75
SPEED OF SOUND 24, 25
SPLITTER 64
SQUARE WAVE 57
STAFF 116
STAGE 19, 36
STANDARD 65, 66
STANDARD VIDEO 47
STANDARDS 50
STAR WARS 41
STATION 19, 26, 47
STEREO 7-9, 17, 19, 23, 30, 33-36, 39, 42, 45, 54, 61, 82, 94, 101
STEREO COMPONENT 18
STEREO EFFECT 24, 25, 38, 57, 98
STEREO HEADPHONES 100
STEREO HI-FI 33, 93, 101
STEREO HI-FI SYSTEM 34
STEREO HI-FI VCR 34
STEREO IMAGE 8, 115
STEREO IMAGING 28, 38, 53
STEREO MECHANICS 24
STEREO MODE 46
STEREO PAIR SETUP 94
STEREO RECEIVER 13, 61, 65
STEREO SET 8
STEREO SIGNAL 46
STEREO SPEAKERS 37, 109
STEREO SYSTEM 9, 11, 27, 38, 76, 88, 89, 109, 111
STEREO TV SIGNALS 46
STEREOPHONIC 8
STEREOPHONIC EFFECT 42, 89
STEREOPHONIC EQUIPMENT 19
STEREOPHONIC TECHNOLOGY 19
STEREOPHONICS 19, 33, 34, 40
STICK 21
STORAGE 30
STORES 8, 113, 116-118
STUDIO 38
STUDIO MAGAZINES 103
SUB/SAT 94
SUB/SAT COMBO 10, 109, 121
SUB/SAT SETUP 97

SUBWOOFER 10, 14, 16, 19, 26, 44, 48, 49, 68, 92, 94-97, 99, 109, 115, 122, 131
SUBWOOFER AND SATELLITE COMBINATION (SUB/SAT) 94
SUBWOOFER CHANNEL 44
SUBWOOFER HOOKUP 131
SUBWOOFER SIGNAL 16, 44
SUBWOOFER SIGNAL LINE 48
SUBWOOFER/SATELLITE (SUB/SAT) 10
SUBWOOFER/SATELLITE COMBO 48
SURFACES 38
SURROUND 67
SURROUND CHANNEL 43, 44, 49
SURROUND EQUIPMENT 10
SURROUND MATRIX 45
SURROUND SIGNALS 12
SURROUND SOUND 8-10, 12, 18, 38, 39, 41-45, 61, 66, 67, 93, 96, 97, 100, 107, 109
SURROUND SOUND DECODERS 10, 65
SURROUND SOUND IMAGE 94
SURROUND SOUND MODES 120
SURROUND SOUND MOVIE TRACKS 46
SURROUND SOUND PACKAGE 119
SURROUND SOUND RECEIVER 43, 68
SURROUND SOUND SETUP 96
SURROUND SOUND SIGNAL SOURCE 47
SURROUND SOUND SPEAKERS 58, 96, 99
SURROUND SOUND STANDARD 50
SURROUND SOUND STEREO 46
SURROUND SOUND SYSTEM 8, 11, 34, 41, 94, 109, 112
SURROUND SOUND TECHNOLOGY 18, 50
SURROUND SPEAKERS 46, 48, 49, 99, 109
SURROUND TECHNOLOGY 39
SURROUNDING AIR 7
SURROUNDINGS 16
SURTAX 76
SWEET SPOT 42, 44
SWITCH 66, 77
SYSTEM 11, 14, 17, 29, 114, 115, 119
SYSTEM SETUP 15

T

TAPE 30, 36, 73, 74, 75, 81, 121
TAPE COUNTER 76
TAPE DECK 6, 9-11, 14, 53, 54, 72-74, 76, 93, 102, 114, 121, 131, 141
TAPE DECK HOOKUP 131
TAPE MEDIA 72
TAPE PLAYER 100
TAPE RECORDERS 34
TAPE RECORDINGS 75

Index 161

TARGET 116
TECHNICS 114
TECHNOLOGY 40, 71, 72, 111
TELEVISION 47, 49, 63, 99
TERMINOLOGY 16
TEST LISTEN 101
TEST SIGNAL GENERATOR 99
TEXTURE 19
THEATER 18
THREE-DIMENSIONAL (3D) 8
THREE-PIECE SYSTEM 94
THRESHOLD OF HEARING 26
THX 16, 41, 50
THX ENGINEERS 50
THX EQUIPMENT 50
THX LOGO 50
THX STANDARDS 50
THX-CERTIFIED COMPONENT 50
THX-CERTIFIED HOME THEATER 50
TIMBER 29
TIME DELAY 24
TIME INTERVAL 80, 106
TONAL MUSICAL QUALITY 29
TONE 92, 107
TONE ARM 84
TONE CONTROL 101
TOSLINK CABLE 124, 128
TOTAL WATTAGE 58
TOWER SPEAKERS 43, 48
TOWER-TYPE CABINET 94
TRACKS 8, 12, 19, 36, 38
TRAIN 26
TRANSDUCER 19, 29, 30, 87, 89, 102
TRANSIENTS 29
TRANSISTOR MODELS 56
TRANSISTORS 5, 51, 55, 56, 68, 77
TRANSMISSION 16
TRANSMITTER 101
TRAVELING DISTANCES 24
TREBLE 7, 29
TREBLE CONTROL 53
TREBLE DRIVER 26
TREBLE FREQUENCIES 53
TREBLE RANGE 100
TREBLE-HIGH 111
TUBES 56
TUNER 6, 11, 13, 16-19, 47, 54, 61-64, 68, 93
TURNTABLE 6, 7, 11, 81, 84, 85, 133
TURNTABLE EQUIPMENT 84
TURNTABLE HOOKUP 133
TURNTABLE NEEDLES 7
TV 12, 97, 128
TV MONITORS 41
TV SET 49
TV SHOWS 45, 46
TV SOUND 111
TV STATION 62

TV TUNERS 54, 64, 67
TV/FM ANTENNA 64
TWEETER 18, 26, 29, 92, 94, 95, 98, 115
TWIN TOWERS 97
TWISTER 96
TWO-CHANNEL 42
TWO-CHANNEL SPEAKER SYSTEM 38
TWO-CHANNEL STEREO 38, 43
TWO-CHANNEL STEREO TECHNOLOGY 38
TWO-STATE COMPONENTS 17
TYPEWRITERS 71

U

UNIDIRECTIONAL MICROPHONE 102
UNIT 15
UNIVERSAL REMOTE CONTROL 66
UPGRADE 108
UPGRADE PLAN 16
USA 17
USED EQUIPMENT 68

V

VACUUM 22
VACUUM TUBE AMP 56
VACUUM TUBES 51, 55
VAN HALEN 81
VARIABLE LEVEL OUTPUTS 54
VARIABLE SPEED PLAY 142
VARYING SIGNAL 17
VCR HOOKUP 133
VCRS 34, 46, 47, 120, 133
VENEER 90
VERBAL MEMOIRS 102
VHS TAPE 46
VIBRATING OBJECT 22
VIBRATIONS 21, 30
VIDEO 12, 17, 46
VIDEO APPLICATIONS 17
VIDEO CABLES 124
VIDEO CD COMPATIBLE 142
VIDEO DISCS 82
VIDEO EQUIPMENT 16
VIDEO GAME 107
VIDEO GAME SOUNDTRACK 109
VIDEO I/OS 67
VIDEO IN/OUT 127
VIDEO INFORMATION 16
VIDEO PATCH CORD 125
VIDEO PICTURE 41
VIDEO SIGNALS 66
VIDEO SOURCES 61
VIDEOTAPE 46
VIEWER 41
VIEWING ROOM 43
VINYL 9, 81, 85, 90
VINYL DISC 84

VIOLIN 58
VIRTUAL SOUND 44, 48
VOLACLIST 36
VOICE SIGNAL 52
VOICES 26, 35, 44, 108
VOLTAGE 16, 79
VOLTAGE LEVELS 17, 78
VOLTAGE MEASUREMENT 79
VOLTAGE READING 79
VOLTAGE VALUES 106
VOLUME 7, 24, 25, 31, 49, 52-54, 57, 58, 75,
    107, 111, 113, 115
VOLUME CONTROL 53, 54, 111, 141, 142

W

WALKMAN 100, 101
WALL 64
WALMART 114, 116
WARP 66
WARRANTY 12, 113, 118
WATER 22
WATTAGE 19, 47-49, 54, 55, 58
WATTAGE RATING 55, 99
WATTS 19, 55, 57
WATTS PER CHANNEL 48, 58
WAVE 16, 18, 21-25, 52, 73, 78, 80, 93
WAVELENGTH 25
WEAK SIGNAL 30
WEAKENED SIGNAL 64
WEBSITES 50, 112, 113
WHISPER 31
WINDOWS MACHINES 108
WIRE 5, 64, 96, 123, 125
WIRING 96, 98
WISH LIST 119
WOOD 95
WOOFER 18, 26, 92, 94
WRITABLE CD 72

Y

YAMAHA 114
YELL 31

Z

ZEROS 77

A Bell Atlantic Company

Howard W. Sams

Your Technology Connection to the Future!

Now You Can Visit Howard W. Sams & Company On-Line:
http://www.hwsams.com

Gain Easy Access to:

- The **PROMPT Publications** catalog, for information on our *Latest Book Releases*.
- The **PHOTOFACT Annual Index**.
- Information on Howard W. Sams' Latest Products.
- *AND MORE!*

PROMPT®
PUBLICATIONS

CALL 1-800-428-7267 TODAY FOR THE NAME OF
YOUR NEAREST PROMPT PUBLICATIONS DISTRIBUTOR

Complete Guide to Video
John Adams

The Video Hacker's Handbook
Carl Bergquist

Complete Guide to Video explains video technology in an easy-to-understand language. It outlines the common components of modern audio/video equipment and gives details and features of the newest gadgets. This book was designed to help you with your shopping choices and to answer questions about video technology that the average salesperson may not know.

An invaluable reference guide, *Complete Guide to Video* will help you set up the right components for your surroundings and will give you the information you need to get the most out of your video equipment.

Geared toward electronic hobbyists and technicians interested in experimenting with the multiple facets of video technology, *The Video Hacker's Handbook* features projects never seen before in book form. Video theory and project information is presented in a practical and easy-to-understand fashion, allowing you to not only learn how video technology came to be so important in today's world, but also how to incorporate this knowledge into projects of your own design. In addition to the hands-on construction projects, the text covers existing video devices useful in this area of technology plus a little history surrounding television and video relay systems.

Video Technology
256 pages - Paperback - 7-3/8 x 9-1/4"
ISBN: 0-7906-1123-6 - Sams: 61123
$24.95 - November 1997

Video Technology
336 pages - Paperback - 7-3/8 x 9-1/4"
ISBN: 0-7906-1126-0 - Sams: 61126
$24.95 - September 1997

CALL 1-800-428-7267 TODAY FOR THE NAME OF YOUR NEAREST PROMPT PUBLICATIONS DISTRIBUTOR

PROMPT PUBLICATIONS

Desktop Digital Video
Ron Grebler

Desktop Digital Video is for those people who have a good understanding of personal computers and want to learn how video (and digital video) fits into the bigger picture. This book will introduce you to the essentials of video engineering, and to the intricacies and intimacies of digital technology. It examines the hardware involved, then explores the variety of different software applications and how to utilize them practically. Best of all, *Desktop Digital Video* will guide you through the development of your own customized digital video system. Topics covered include the video signal, digital video theory, digital video editing programs, hardware, digital video software and much more.

Video Technology
225 pages - Paperback - 7-3/8 x 9-1/4"
ISBN: 0-7906-1095-7 - Sams: 61095
$29.95 - June 1997

TV Video Systems
L.W. Pena & Brent A. Pena

Knowing which video programming source to choose, and knowing what to do with it once you have it, can seem overwhelming. Covering standard hard-wired cable, large-dish satellite systems, and DSS, *TV Video Systems* explains the different systems, how they are installed, their advantages and disadvantages, and how to troubleshoot problems. This book presents easy-to-understand information and illustrations covering installation instructions, home options, apartment options, detecting and repairing problems, and more. The in-depth chapters guide you through your TV video project to a successful conclusion.

Video Technology
124 pages - Paperback - 6 x 9"
ISBN: 0-7906-1082-5 - Sams: 61082
$14.95 ($20.95 Canada) - June 1996

CALL 1-800-428-7267 TODAY FOR THE NAME OF YOUR NEAREST PROMPT PUBLICATIONS DISTRIBUTOR

PROMPT® PUBLICATIONS

RadioScience Observing
Volume 1
Joseph J. Carr

Among the hottest topics right now are those related to radio: radio astronomy, propagation studies, whistler and spheric hunting, searching for solar flares using VLF radio, and related subjects. Author Joseph Carr lists all of these topics under the term RadioScience Observing: a term he has coined to cover the entire field.

In *RadioScience Observing*, you will find chapters on all of these topics and more. The main focus on the book is the amateur scientist who has a special interest in radio. It is also designed to appeal to amateur radio enthusiasts, shortwave listeners, scanner band receiver owners, and other radio hobbyists. *RadioScience Observing* is a useful adjunct to an already thriving hobby.

CD-ROM Included!

Security Systems for Your Home & Automobile
by Gordon McComb

Security Systems is about making homes safer places to live and protecting cars from vandals and thieves. It is not only a buyer's guide to help readers select the right kind of alarm system for their home and auto, it also shows them how to install the various components. Learning to design, install, and use alarm systems saves a great deal of money, but it also allows people to learn the ins and outs of the system so that it can be used more effectively. This book is divided into eight chapters, including home security basics, warning devices, sensors, control units, remote paging automotive systems, and case histories.

Gordon McComb has written over 35 books and 1,000 magazine articles which have appeared in such publications as *Omni* and *PC World*. In addition, he is the coauthor of PROMPT® Publication's *Speakers for Your Home and Auto*.

Electronics Technology
432 pages + Paperback + 7-3/8 x 9-1/4"
ISBN: 0-7906-1127-9 + Sams: 61127
$29.95 + January 1998

Projects
130 pages + Paperback + 6 x 9"
ISBN: 0-7906-1054-X + Sams: 61054
$16.95 + July 1994

**CALL 1-800-428-7267 TODAY FOR THE NAME OF
YOUR NEAREST PROMPT PUBLICATIONS DISTRIBUTOR**

PROMPT PUBLICATIONS

PC Hardware Projects Volume 1
James "J.J." Barbarello

Now you can create your own PC-based digital design workstation! Using commonly available components and standard construction techniques, you can build some key tools to troubleshoot digital circuits and test your printer, fax, modem, and other multiconductor cables.

This book will guide you through the construction of a channel logic analyzer, and a multipath continuity tester. You will also be able to combine the projects with an appropriate power supply and a prototyping solderless breadboard system into a single digital workstation interface!

PC Hardware Projects, Volume 1, guides you through every step of the construction process and shows you how to check your progress.

PROJECT SOFTWARE DISK INCLUDED!

Computer Technology
256 pages - Paperback - 7-3/8 x 9-1/4"
ISBN: 0-7906-1104-X - Sams: 61104
$24.95 - Feb. 1997

PC Hardware Projects Volume 2
James "J.J." Barbarello

PC Hardware Projects, Volume 2, discusses stepper motors, how they differ from conventional and servo motors, and how to control them. It investigates different methods to control stepper motors, and provides you with circuitry for a dedicated IC controller and a discrete component hardware controller.

Then, this book guides you through every step of constructing an automated, PC-controlled drilling machine. You'll then walk through an actual design layout, creating a PC design and board. Finally, you'll see how the drill data is determined from the layout and drill the PCB. With the help of the information and the data file disk included, you'll have transformed your PC into your very won PCB fabrication house!

PROJECT SOFTWARE DISK INCLUDED!

Computer Technology
256 pages - Paperback - 7-3/8 x 9-1/4"
ISBN: 0-7906-1109-0 - Sams: 61109
$24.95 - May 1997

CALL 1-800-428-7267 TODAY FOR THE NAME OF YOUR NEAREST PROMPT PUBLICATIONS DISTRIBUTOR

PROMPT PUBLICATIONS

Real-World Interfacing With Your PC
Second Edition
James "J.J." Barbarello

Real-World Interfacing With Your PC, 2nd Edition, provides you with all the information you need to use your PC's parallel port as a gateway to real-world electronic interfacing.

This book provides you with a basic understanding of writing software to control the hardware. It also walks you through an actual project, from design to construction. If you're not an experienced electronics builder, there's a chapter on project construction techniques and a project builder's checklist. *Real-World Interfacing With Your PC* even gets you on the way to your next project, and recommends a "starter" inventory of electronic parts and how to market your best ideas.

PROJECT SOFTWARE DISK INCLUDED!

Computer Technology
224 pages - Paperback - 7-3/8 x 9-1/4"
ISBN: 0-7906-1145-7 - Sams: 61145
$24.95 - July 1996

Alternative Energy
Mark E. Hazen

This book is designed to introduce readers to the many different forms of energy mankind has learned to put to use. Generally, energy sources are harnessed for the purpose of producing electricity. This process relies on transducers to transform energy from one form into another. *Alternative Energy* will not only address transducers and the five most common sources of energy that can be converted to electricity, it will also explore solar energy, the harnessing of the wind for energy, geothermal energy, and nuclear energy.

This book is designed to be an introduction to energy and alternate sources of electricity. Each of the nine chapters are followed by questions to test comprehension, making it ideal for students and teachers alike. In addition, listings of World Wide Web sites are included so that readers can learn more about alternative energy and the organizations devoted to it.

Professional Reference
320 pages - Paperback - 7-3/8 x 9-1/4"
ISBN: 0-7906-1079-5 - Sams: 61079
$18.95 ($25.95 Canada) - October 1996

CALL 1-800-428-7267 TODAY FOR THE NAME OF YOUR NEAREST PROMPT PUBLICATIONS DISTRIBUTOR

PROMPT PUBLICATIONS

Electronic Circuit Guidebook Volume 1, Sensors
Joseph J. Carr

Most sensors are inherently analog in nature, so their outputs are not usable by the digital computer. Even if the sensor is supposedly a digital output design, it is likely that an inherently analog process is paired with an analog-to-digital converter. In *Electronic Circuit Guidebook, Volume 1: Sensors*, you will find information you need about typical sensors, along with a large amount of information about analog sensor circuitry. Amplifier circuits are especially well covered, along with differential amplifiers, analog signal processing circuits and more. This book is intentionally kept practical in outlook. Some topics covered include electronics signals and noise, measurement, sensors and instruments, instrument design rules, sensor interfaces, analog amplifiers, and sensor resolution improvement techniques.

Electronic Theory
340 pages + Paperback + 7-3/8 x 9-1/4"
ISBN: 0-7906-1098-1 + Sams: 61098
$24.95 + April 1997

Electronic Circuit Guidebook Volume 2, IC Timers
Joseph J. Carr

Timer circuits used to be a lot of trouble to build and tame for several reasons. One major reason was the fact that DC power supply variations would cause a frequency shift or slow drift. Part I of this book is organized to demonstrate the theory of how various timers work. This is done by way of an introduction to resistor-capacitor circuits, and in-depth chapters on various TTL and CMOS digital IC devices. Part II presents a variety of different circuits and projects. Some of the circuits include: analog audio frequency meter, one-second timer/flasher, relay and optoisolator drivers, two-phase digital clock and more. *Electronic Circuit Guidebook, Volume 2: IC Timers* will teach you enough that you will not only be able to rework and modify the circuits covered here, but also design a few of your own.

Electronics Technology
240 pages + Paperback + 7-3/8 x 9-1/4"
ISBN: 0-7906-1106-6 + Sams: 61106
$24.95 + August 1997

CALL 1-800-428-7267 TODAY FOR THE NAME OF YOUR NEAREST PROMPT PUBLICATIONS DISTRIBUTOR

Electronic Circuit Guidebook Volume 3, Op Amps
Joseph J. Carr

The operational amplifier is the most commonly used linear IC amplifier in the world. The range of applications for the op amp is truly awesome – it has become a mainstay of audio, communications, TV, broadcasting, instrumentation, control, and measurement circuits. Third in a series covering electronic instrumentation and circuitry, *Electronic Circuit Guidebook, Volume 3: Op Amps* is design to give you some insight into how practical linear IC amplifiers work in actual real-life circuits. Because of their widespread popularity, operational amplifiers figure heavily in this book, though other types of amplifiers are not overlooked. This book allows you to design and configure your own circuits, and is intended to be a practical workshop aid. Some of the topics covered in detail include linear IC amplifiers, ideal operational amplifiers, instrumentation amplifiers, isolation amplifiers, active analog filter circuits, waveform generators, and many more.

Electronics Technology
273 pages + Paperback + 7-3/8 x 9-1/4"
ISBN: 0-7906-1131-7 + Sams: 61131
$24.95 + August 1997

Electronic Circuit Guidebook Volume 4, Electro-Optics
Joseph J. Carr

Electronic Circuit Guidebook, Volume 4: Electro-Optics is mostly about E-O sensors — those electronic transducers that convert light waves into a proportional voltage, current, or resistance. The coverage of the sensors is wide enough to allow you to understand the physics behind the theory of operation of the device, and also the circuits used to make these sensors into useful devices. This book examines the photoelectric effect, photoconductivity, photovoltaics, and PN junction photodiodes and phototransistors. Also examined is the operation of lenses, mirrors, prisms, and other optical elements keyed to light physics.

Electronic Circuit Guidebook, Volume 4: Electro-Optics is intended to teach the physics and operation of E-O devices, then proceed to circuits and methods for actual application of the devices in real situations.

Electronics Technology
416 pages + Paperback + 7-3/8 x 9-1/4"
ISBN: 0-7906-1132-5 + Sams: 61132
$29.95 + October 1997

CALL 1-800-428-7267 TODAY FOR THE NAME OF YOUR NEAREST PROMPT PUBLICATIONS DISTRIBUTOR

PROMPT PUBLICATIONS

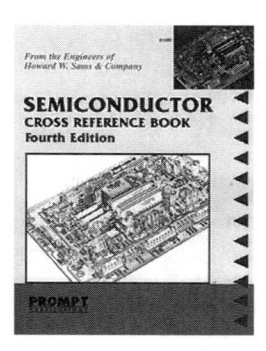

Semiconductor Cross Reference Book Fourth Edition
Howard W. Sams & Company

This newly revised and updated reference book is the most comprehensive guide to replacement data available for engineers, technicians, and those who work with semiconductors. With more than 490,000 part numbers, type numbers, and other identifying numbers listed, technicians will have no problem locating the replacement or substitution information needed. There is not another book on the market that can rival the breadth and reliability of information available in the fourth edition of the *Semiconductor Cross Reference Book*.

Professional Reference
688 pages - Paperback - 8-1/2 x 11"
ISBN: 0-7906-1080-9 - Sams: 61080
$24.95 ($33.95 Canada) - August 1996

IC Cross Reference Book Second Edition
Howard W. Sams & Company

The engineering staff of Howard W. Sams & Company assembled the *IC Cross Reference Book* to help readers find replacements or substitutions for more than 35,000 ICs and modules. It is an easy-to-use cross reference guide and includes part numbers for the United States, Europe, and the Far East. This reference book was compiled from manufacturers' data and from the analysis of consumer electronics devices for PHOTOFACT® service data, which has been relied upon since 1946 by service technicians worldwide.

Professional Reference
192 pages - Paperback - 8-1/2 x 11"
ISBN: 0-7906-1096-5 - Sams: 61096
$19.95 ($26.99 Canada) - November 1996

CALL 1-800-428-7267 TODAY FOR THE NAME OF YOUR NEAREST PROMPT PUBLICATIONS DISTRIBUTOR

PROMPT PUBLICATIONS

The Component Identifier and Source Book
Victor Meeldijk

Because interface designs are often reverse engineered using component data or block diagrams that list only part numbers, technicians are often forced to search for replacement parts armed only with manufacturer logos and part numbers.

This source book was written to assist technicians and system designers in identifying components from prefixes and logos, as well as find sources for various types of microcircuits and other components. There is not another book on the market that lists as many manufacturers of such diverse electronic components.

Tube Substitution Handbook
William Smith & Barry Buchanan

The most accurate, up-to-date guide available, the *Tube Substitution Handbook* is useful to antique radio buffs, old car enthusiasts, and collectors of vintage ham radio equipment. In addition, marine operators, microwave repair technicians, and TV and radio technicians will find the *Handbook* to be an invaluable reference tool.

The *Tube Substitution Handbook* is divided into three sections, each preceded by specific instructions. These sections are vacuum tubes, picture tubes, and tube basing diagrams.

Professional Reference
384 pages - Paperback - 8-1/2 x 11"
ISBN: 0-7906-1088-4 - Sams: 61088
$24.95 ($33.95 Canada) - November 1996

Professional Reference
149 pages - Paperback - 6 x 9"
ISBN: 0-7906-1036-1 - Sams: 61036
$16.95 ($22.99 Canada) - March 1995

CALL 1-800-428-7267 TODAY FOR THE NAME OF YOUR NEAREST PROMPT PUBLICATIONS DISTRIBUTOR

PROMPT PUBLICATIONS

The Microcontroller Beginner's Handbook
Lawrence A. Duarte

Microcontrollers are found everywhere — microwaves, coffee makers, telephones, cars, toys, TVs, washers and dryers. This book will bring information to the reader on how to understand, repair, or design a device incorporating a microcontroller. *The Microcontroller Beginner's Handbook* examines many important elements of microcontroller use, including such industrial considerations as price vs. performance and firmware. A wide variety of third-party development tools is also covered, both hardware and software, with emphasis placed on new project design. This book not only teaches readers with a basic knowledge of electronics how to design microcontroller projects, it greatly enhances the reader's ability to repair such devices. Lawrence A. Duarte is an electrical engineer for Display Devices, Inc. In this capacity, and as a consultant for other companies in the Denver area, he designs microcontroller applications.

Electronic Theory
240 pages - Paperback - 7-3/8 x 9-1/4"
ISBN: 0-7906-1083-3 - Sams: 61083
$18.95 ($25.95 Canada) - July 1996

Schematic Diagrams
J. Richard Johnson

Step by step, *Schematic Diagrams* shows the reader how to recognize schematic symbols and determine their uses and functions in diagrams. Readers will also learn how to design, maintain, and repair electronic equipment as this book takes them logically through the fundamentals of schematic diagrams. Subjects covered include component symbols and diagram formation, functional sequence and block diagrams, power supplies, audio system diagrams, interpreting television receiver diagrams, and computer diagrams. *Schematic Diagrams* is an invaluable instructional tool for students and hobbyists, and an excellent guide for technicians.

Electronic Theory
196 pages - Paperback - 6 x 9"
ISBN: 0-7906-1059-0 - Sams: 61059
$16.95 ($22.99 Canada) - October 1994

CALL 1-800-428-7267 TODAY FOR THE NAME OF YOUR NEAREST PROMPT PUBLICATIONS DISTRIBUTOR

PROMPT® PUBLICATIONS

The Howard W. Sams Troubleshooting & Repair Guide to TV
Howard W. Sams & Company

The Howard W. Sams Troubleshooting & Repair Guide to TV is the most complete and up-to-date television repair book available. Included in its more than 300 pages is complete repair information for all makes of TVs, timesaving features that even the pros don't know, comprehensive basic electronics information, and extensive coverage of common TV symptoms.

This repair guide is completely illustrated with useful photos, schematics, graphs, and flowcharts. It covers audio, video, technician safety, test equipment, power supplies, picture-in-picture, and much more. *The Howard W. Sams Troubleshooting & Repair Guide to TV* was written, illustrated, and assembled by the engineers and technicians of Howard W. Sams & Company.

The In-Home VCR Mechanical Repair & Cleaning Guide
Curt Reeder

Like any machine that is used in the home or office, a VCR requires minimal service to keep it functioning well and for a long time. However, a technical or electrical engineering degree is not required to begin regular maintenance on a VCR. *The In-Home VCR Mechanical Repair & Cleaning Guide* shows readers the tricks and secrets of VCR maintenance using just a few small hand tools, such as tweezers and a power screwdriver.

This book is also geared toward entrepreneurs who may consider starting a new VCR service business of their own. The vast information contained in this guide gives a firm foundation on which to create a personal niche in this unique service business. This book is compiled from the most frequent VCR malfunctions Curt Reeder has encountered in the six years he has operated his in-home VCR repair and cleaning service.

Video Technology
384 pages - Paperback - 8-1/2 x 11"
ISBN: 0-7906-1077-9 - Sams: 61077
$29.95 ($39.95 Canada) - June 1996

Video Technology
222 pages - Paperback - 8-3/8 x 10-7/8"
ISBN: 0-7906-1076-0 - Sams: 61076
$19.95 ($26.99 Canada) - April 1996

CALL 1-800-428-7267 TODAY FOR THE NAME OF YOUR NEAREST PROMPT PUBLICATIONS DISTRIBUTOR

PROMPT PUBLICATIONS

Howard W. Sams Complete VCR Troubleshooting & Repair
Joe Desposito & Kevin Garabedian

Complete VCR Troubleshooting and Repair contains sound VCR troubleshooting procedures beginning with an examination of the external parts of the VCR, then narrowing the view to gears, springs, pulleys, belts, and other mechanical parts. This book also shows how to troubleshoot tuner/demodulator circuits, audio and video circuits, display controls, servo systems, video heads, TV/VCR combination models, and more.

This book also contains nine VCR case studies, each focusing on a particular model of VCR with a specific problem. The case studies guide you through the repair from start to finish, using written instruction, helpful photographs, and Howard W. Sams' own *VCRfacts*® schematics.

Video Technology
184 pages - Paperback - 8-1/2 x 11"
ISBN: 0-7906-1102-3 - Sams: 61102
$29.95 - March 1997

Howard W. Sams Computer Monitor Troubleshooting & Repair
Joe Desposito & Kevin Garabedian

Computer Monitor Troubleshooting & Repair makes it easier for any technician, hobbyist or computer owner to successfully repair dysfunctional monitors. Learn the basics of computer monitors with chapters on tools and test equipment, monitor types, special procedures, how to find a problem and how to repair faults in the CRT. Other chapters show how to troubleshoot circuits such as power supply, high voltage, vertical, sync and video.

This book also contains six case studies which focus on a specific model of computer monitor. Using carefully written instructions and helpful photographs, the case studies guide you through the repair of a particular problem from start to finish. The problems addressed include a completely dead monitor, dysfunctional horizontal width control, bad resistors, dim display and more.

Troubleshooting & Repair
308 pages - Paperback - 8-1/2 x 11"
ISBN: 0-7906-1100-7 - Sams: 61100
$29.95 - July 1997

CALL 1-800-428-7267 TODAY FOR THE NAME OF YOUR NEAREST PROMPT PUBLICATIONS DISTRIBUTOR